Marketing Grassfed Products Profitably

Marketing Grassfed Products Profitably

by

Carolyn Nation

preface by

Allan Nation

GREEN PARK PRESS

A division of Mississippi Valley Publishing Corp.
Ridgeland, Mississippi

First printing January 2011

Library of Congress Cataloging-in-Publication Data

Nation, Carolyn.
Marketing grassfed products profitably / by Carolyn Nation ; preface by Allan Nation.
p. cm.
Includes bibliographical references and index.
ISBN 0-9721597-6-2 (alk. paper)
1. Meat--Marketing. 2. Milk trade. 3. Direct selling. I. Nation, Allan. II. Title.
SF 113.3.N39 2011
664'.900688--dc22
2010039386

Cover photos by Carolyn Nation
Cover design by Steve Erickson, Madison, MS

Manufactured in the United States.
This book is printed on recycled paper.

Table of Contents

Preface — Why direct marketing should be women's work...9
Introduction — Putting together the puzzle pieces...14

First Things First

Why me?...19
Women are born marketers...23
The importance of setting goals...28
Creating life planning goals...33
Making your goals three dimensional...36

What Are You Selling?

Quality is job one...42
Remarkable products...44
Providing value with values...48
Authenticity really matters...53
Be the brand...57
Selling benefits..63.
Naming names...68

How Much Should You Charge?

Pricing...75
Responding to price objections...82
Offering purchases for penny pinchers...86

Marketing — Make A Plan

Creating a marketing plan...93
21 questions about your competition...96
Touting your message...100

How To Sell — Putting The Customer First

Who are your customers?...107
20 questions about your customers...111
Listening...115
How to get customers to sell themselves...119
Viva la difference!...123
Guidelines for men selling to women...128
Guidelines for women selling to men...132
How to make meangingful connections...136
What makes "feel good" sales...140
Multitasking marketing...145
Helping stressed out shoppers...149
Stressed out sellers...153
Putting senior customers first...157
Theatre, making food fun...162
Luring customers with samples...167
Reading between the lines to close the sale...171
Anger management — how to handle criticism...177
Educating your customers...181
Building community...188

Where To Sell

Farmers' markets...193
Practical issues of farmers' markets...196
Making your booth stand out at the farmers' market...200
Meat displays...203
Professionalism...207
Making farmers' markets fun...212
Load the wagon...217
Value added products...221
Service...226
Celebrating special occasions and milestones...230
Beyond farmers' markets — the farm store...235
Location...241
Organizing a Buying Club...245
Beyond the farm store...252

Convincing chefs to go grassfed...259
The farm as destination...264
Farm tours, how tos...271
The farm stay...274
Is there a festival in your future?...278

Professionalism

Database marketing...287
Giving thanks...291
Appearance...293
Pretend to be your customer...298
Selling by coconut telegraph...302
Public relations...305
Advertising...310
Marketing messages: good and bad...315
Logos...319
Brochures...322
Catalogs...325
Newsletters...330
Internet marketing...333
Invoices and order forms...343
Compiling a Farm Manual...347
Copycats and niches...351
Change...358

Bibliography...358
Index...362
About the author...364

NOTE: Just as in nature, things change. Many of the producer examples described in this book may not be the same today as when the book was written. However, these marketing suggestions are timeless, limited only to the vision and imagination of savvy direct marketers who will adapt them for their own business.

Preface
Why direct marketing should be women's work

I was at a large farmers' market in a major Northeastern city. As I typically do, I scouted around to see if any exhibitors had grassfed products for sale.

And yes, there were two. One was for grassfed beef and another for grassfed cheese.

In the grassfed beef booth a man stood behind a table with a large plastic cooler set upon it.

A small hand-lettered sign simply said, "Grassfed Beef For Sale."

His arms were crossed in front of him and he had a cowboy hat pulled down to his nose so that you couldn't see his eyes.

His whole demeanor and body language shouted, "Don't you dare come over here and try to talk to me!"

And, guess what? Nobody did.

I did not see a single customer ever go up to the booth in the hour or so I was there.

This was really painful to watch.

And, what hurt me the most was that this was not some recent startup, but a well-known grassfed beef company.

Now, I am guessing this guy was a hand that had been "sent" by his boss because the regular marketing person was not available.

Hopefully, this was the case. Otherwise, this beef

company will soon be history if the farmers' market is important to its sales.

In contrast, at the grassfed cheese booth there were lots of signs, posters and photos discussing and displaying the farm.

There were two attractive young women manning the booth who appeared to really enjoy talking about cheese.

The male cheesemaker hovered in the background, available in case anyone had a technical question, but otherwise he let the two women do all the talking.

Apparently this worked as there was always a line of customers and sales were brisk.

Later I ruminated over what I had just witnessed.

I thought to myself, "If there was one thing I could change in the beef booth to increase sales, what would it be?"

Better signage?

A better product display?

It suddenly hit me.

The best way to increase this ranch's sales would be to exchange the man in the booth for a woman!

One of the things I have long noted but never really thought about is that the family-owned operations where the marketing is handled by the wife are almost invariably far more successful that those where it is handled by the husband.

This is true even when the marketing has been dumped on wives by their overwhelmed husbands and they have no formal marketing training or prior experience.

I got to thinking about this and it hit me!

As marketing guru Faith Popcorn once noted, "Men may eat steak, but women buy it."

According to management consultant Tom Peters, women buy 83% of all consumer purchases. In home-cooked food products, this is probably closer to 94%.

And, guess what?

Women can sell to women better than men can because they know how a woman thinks.

Here's the root problem for males.

Men instinctively size up everyone they meet as to where they are in the hierarchal order.

"Should I have more respect for you, or are you somebody that's south of me?" men think.

Needless to say, if you are one down you are going to be treated with a tiny bit of disdain and disrespect.

No doubt every man has felt this from some snooty sales guy in an upscale retail environment.

Guess what? Most men see **all** women as one down.

And, do women feel the condescension? Yes!

Women have their antennae out all the time and most of us male farmers and ranchers are not exactly sensitive, new-age, guys.

A female *Stockman Grass Farmer* reader recently related this story to me:

She was waiting her turn at a booth at a Southeastern city's farmers' market and a woman in front of her directed a technical question about pasturing to the farmer's wife who was waiting on her.

The farmer's wife started to answer but her husband pushed her aside and answered the woman's question.

The woman thanked him and walked away without buying anything.

As did every other woman who had been waiting to buy.

She said that any man who showed that much disrespect for his wife in public would not get her business and apparently all the other women waiting agreed.

She said the correct protocol is the one I saw in the cheese booth whereby the man is available if the female salesperson can't answer the question, but that he only speaks after having been invited to do so.

Judith Tingley in *GenderSell: How to Sell to the Opposite Sex* said women tend to see male salespeople as "technically knowledgeable" and "assertive," they get "straight to the point."

Women also see them as "pushy," "condescending," and "insensitive to women's needs."

Ouch!

When men buy they like speed in the transaction with as little conversation as possible.

A simple sign saying "Grassfed Beef" is probably enough product information for a man but it isn't for a woman.

When women buy they want context before content.

Prolific signage, photos, philosophy and anything else personal you can hang on your booth's walls is important to women consumers because it tells *who* you are.

This seeking of context is also why women tend to be much more verbal in their initial buying experience than men.

"He simply wants the transaction to take place; she's interested in establishing the relationship," Faith Popcorn said.

Tom Peters said that women are harder to sell initially than men but are very loyal once sold.

Men, on the other hand, are just the opposite, easy to sell initially, but hard to keep.

The average woman is also a much better "sneezer" (one who spreads the word) than a male customer, which is critical in keeping marketing costs low by spreading the good word about things they love.

Peters said give a man an exceptional product and service and he will tell 2.6 other people.

Give a woman the same thing and she will tell 21 other people.

Women are also much more receptive to the health angle of alternative foods than men. However, several women consumers have told me this concern can be vetoed by a husband if the product does not taste good.

There have been whole books written about gender differences and their effect on marketing, however, I hope I have shown with these few examples why a woman probably should be doing the ranch's consumer marketing.

While this book is written by a woman with a female

audience in mind, the marketing principles are universal.

With this book, those guys among us can learn to be better marketers as well. And maybe we'll have a better understanding of our spouses too.

Allan Nation
Editor, *Stockman Grass Farmer*
And Carolyn's spouse

Introduction
Putting together the puzzle pieces

Working at the *Stockman Grass Farmer* magazine was not my chosen career. Nor was it the first choice for Glinda Davenport, our partner and managing editor.

Both of us were busily minding our own business when our spouses asked us to try working on "this" and to help out on "that."

Glinda's call to "Help!" came at a crucial time when the heavily in debt magazine was trying to dig itself out of the red.

My involvement was more subtle when I was asked to head up the conference division and later to create and implement the publishing of books for the Green Park Press division.

Both of us are involved in getting the magazine out each month, although you would rarely see a by-line by either of us. Our roles involve attention to all the details - crossing the "t"s and dotting the "i"s in proofreading, managing files, and getting all the pages together and prepped for the printer.

Attention to details is a trait in which women generally excel.

As spouses, we have a vested interest in making the business succeed. We're less likely to seek better employment elsewhere or leave when things get tough, so the ordeal of finding a replacement is less likely to occur.

This insures continuity within the business and avoids lost productivity involved with training someone new. Sometimes we'll work for, well not peanuts exactly, but maybe a Prime grass-finished ribeye.

Allan and I work together in our home. This is a very harmonious situation because we have separate offices, and separate jobs. He calls the jobs he tossed my way "balls."

While I may get his opinion or suggestions from other staff members, I'm totally in charge of the book publishing division. That means no interfering — I mean, second guessing — from Allan.

Book publishing is my baby. I'm empowered to make all the choices and decisions about the books we publish, how they will be published, and what they'll look like.

The key here is that we are working on different aspects of the business, empowered in our own areas.

So what does this have to do with grass farming? A lot!

If you don't have the support of your spouse and shared goals, the journey to success will be harder than it needs to be.

How many of you have had your spouse dump undesirable aspects of the business on you? "Here, Honey, you keep the books. You write all the checks anyway."

Or "I'm busy planning my pasture forage chain. You go to the farmers' market."

Don't panic. Even if you don't at first feel qualified. This is a blessing. You should be involved in the family grass farming operation.

Before starting remember

- Working together works best when you have separate jobs or enterprises.
- Spouses have a vested interest in their family's business.
- Women are good with details.
- Shared goals and support of your spouse equals success.

Even if you feel overwhelmed at first, there are many areas where women are better suited than men. The purpose of this book is to share with you marketing ideas, how to work harmoniously with your spouse, why it's important to make time for family and yourself, and what to do to promote your farm.

First Things First

Why me?

With the majority of North American farms being family owned, it's only natural for women — wives — to be involved.

Never mind if you're a city girl who can barely tell the difference between a bull and cow. Forget about any degrees you might have that aren't ag-related. And just ignore that off-farm job. The day will come when you're asked, or maybe even dragged kicking and screaming into the farm business. And that's when you're most likely to protest, "Why me?"

If you want the farm business to succeed, you need to be involved, not only to share your spouse's vision and dreams, but the failures and more importantly the successes.

This doesn't mean you'll be expected to learn the names of cool-season, warm-season grasses, brassicas and legumes, and when to plant and when they grow. For those aspects you can rely on your husband's skill and expertise in knowing when it's time for a paddock shift. But perhaps he's the strong and silent type who thrives on building fences and figuring out water systems for the animals, but he'd rather let you do the talking when it comes to marketing. And that's where we women shine. It's a universal truth.

Let's look at an example from an entirely different industry, but which follows parallels in the farming business.

Madalene Daniell, my best friend, married into her

husband Steve's business—a used car dealership. After years of watching from the sidelines while raising their two children, Sarah and Breck, and helping with the bookkeeping, Steve turned to her one morning and admitted, "I feel like I'm paddling this canoe upstream as hard as I can and losing ground. I need your help."

It was a pivotal time for Madalene. Her father, an educator and computer whizz, had died three weeks earlier, leaving her haunted by his advice to get involved in her husband's business and computerize. And Steve had watched her, as Booster Club president for her son's high school, raise $60,000 for a new athletics building and set in place a computer tracking system for the athletic students.

The business was in debt, "going belly up faster than I knew," Madalene recalled. Steve predictably resisted the computers and didn't want to spend money for anything he didn't understand. A secretary had been hand posting expenses, but when Madalene took over the books on January 1, she discovered the last posting had been in June. "He didn't like me replacing his little black hand-written record book," she said. "He kept pulling me in and pushing me out. It was the yo-yo of all yo-yos."

She had never wanted to be involved in what she considered "his thing." With a master's degree in dance and a comfortable career teaching postpartum and prenatal exercise classes for a medical clinic, she wasn't trained for his business and had few clues to its complexity. But Steve had admitted he was tapped out enough to ask for her help.

"It's a desperate place," she said, "because otherwise you're not teachable. You're not open to listen or hear what's wrong. You're in denial."

Comparing her business and its income to his, she realized Steve's business had the most to offer as a family income. But it also had the most to lose. Realizing she might, literally, lose her home if she didn't help was the ultimate motivator to get involved.

Why it's important to be involved in your spouse's business:

■ You can't appreciate what he does if you're not involved to some degree. Otherwise, it's easy to take for granted what he does. It's like the difference between being given an inheritance and earning it.

■ In business you need different skillsets, someone you can trust and who cares about the business. You can't hire "ownership." Even if you hire that rare employee, there's no guarantee they'll stay with you.

■ Working together builds appreciation and trust on both sides.

■ It keeps you together as life changes — children get older and move away, but you continue to have a "baby" together (with the business), investing in something that has life and you can watch grow. It helps you grow together rather than apart.

■ You'll share a common vision.

■ Having common values on financial management keeps both of you reigned in on financial decisions and gives you balance.

■ When important decisions need to be made (upgrading computers, marketing, advertising, educational conferences, customer service) the spouse has more valuable input than an employee would.

■ You develop a common work ethic and family values.

■ Recognizing each other's gifting enhances the business and the marriage.

Madalene Daniell, www.daniellmotors.com

One of the first things they did was write off over $8000 in long-overdue accounts receivables just to have a fresh start.

"If the foundation of the business is not good," she said, "you better strip it down to the foundation rather than try to make walls that aren't plumb, plumb. You have to get the foundation right before you can move forward, and if you're planning to go up, you have to go down first."

Today, their family business is successfully growing and evolving. Steve excels at purchasing. Their son, Breck, handles sales and financing. Madalene watches over the books, public relations, and human resources. Daniell Motors was awarded Independent Dealer of the Year award in 2007. And Madalene still fits in time to teach her classes.

"If you bring in your spouse," Madalene said, "you have to accept who that person is and has to offer. And you have to give them license to do what they're supposed to do. Don't try to make me your mother or sister, because I don't iron sheets or like to cook. Don't make me something I'm not."

She noted that if you work together as a team, if you remove one person, as in the movie, "It's A Wonderful Life," the whole town (business) changes. Since opposites usually attract, if two people are alike, one is unnecessary.

Madalene keeps that in mind when hiring employees. "Don't make the person what you want them to be," she suggested. "Find their gifting and put them in a position that represents who they are, not just what you need them to do. Try to find their strength and grow the business based on that. It will make them feel valued if they're doing what they're gifted for. (Our business) is about helping the customer. It's not about selling cars." Or cows. Or anything else.

Women are born marketers

Were you one of those enterprising kids who set up a lemonade stand in the front yard? Did you sell Girl Scout cookies door-to-door? How many of you sweet-talked your relatives and their grown up friends into buying raffle tickets for a school fundraiser or worthy cause?

Girls, and women, are natural born marketers. It's a universal truth.

Maybe the idea of *selling* makes you feel uneasy. Perhaps that's because you've always thought of sales being a man's thing, somewhat unladylike.

That's Old Think—being pushy, coercive, using the hard sell. But let's give the guys some credit. Some are excellent marketers of grassfed products. Still, many (most?) would rather concentrate on the production end of the farm business and leave the selling to you.

New Think suits a woman's natural style—being intuitive, picking up on clues about the buyer's desires and needs, communicating with sensitivity and good humor, being friendly. In other words, being yourself. Remember, Ginger Rogers did everything Fred Astaire did, but she did it backwards and in high heels.

Easy does it. If marketing is a new role to add to your already stacked multi-tasking skills, the first thing to do is relax. Realize you won't become an expert marketer overnight.

It may take awhile to find your rhythm and comfort zone. In a farmers' market setting, if you don't sell out the first few times it's not a sign of your incompetence.

Experts say it takes eight "touches" before someone commits to buy. These "touches" are warm-ups to the purchase. A person has to see the farm's name, recognize that they've seen it before, hear someone talking about it, notice that someone else is interested in the product, focus on the aroma of samples sizzling on a grill, chew on the sample, chat with the person in the booth, think about the experience, and then they're open enough to consider buying. This probably won't all happen in one day.

Just being there, on a regular consistent (persistent) schedule, is your first step in establishing familiarity with you and your products. It shows that your business is here to stay, not a fly-by-night operation like those people who sell paintings of Elvis on velvet from the back of a truck.

Intuition. Rudyard Kipling said, "A woman's guess is much more accurate than a man's certainty."

Women's intuition isn't magical. We're curious and like to talk. Women's friendliness makes it easy for us to strike up a conversation with complete strangers. We ask questions and listen to the answers. Sometimes that might mean reading between the lines when it comes to anticipating the needs of others, or reading body language clues.

Haven't you known without spoken words when that puppy needs to go outside or is hungry?

Perhaps you have a prospective buyer who's intrigued after sampling your lamb kebabs. Chatting with them you sense their buying hesitancy. By asking questions you learn their reluctance stems from the fear of ruining a pricy cut of meat by improper cooking, or they're afraid of something like lamb or a huge brisket because they've never purchased or cooked it before.

Now you have something to work with. You can defuse their fear by suggesting easier things to start with—ground lamb for meatballs for example, with recipes to guide them. Maybe they won't buy your most expensive offering the first time. But little sales open the door to bigger ones once you've established rapport and trust with your customers.

Persuasion. A friend of ours knows her husband hates tuna fish. So, whenever she's had a hard day and wants to go out to dinner, she sets a large can of tuna on the kitchen counter. When her husband comes in, probably equally exhausted from his work day and sees that can of tuna, he always suggests, "Let's go out to eat." She gets what she wants without saying a word.

Similarly, when I'm shopping for clothes, I'm more likely to splurge when someone assists (or gently persuades) me by selecting things off the rack for me. A little flattery helps cinch the deal.

Left alone, I find excuses—funky color, it's not cotton,

Can you pass the test?

1. Do you enjoy talking to people?
2. Are you a good listener?
3. Are you gifted with a memory for recalling others' names?
4. Do you make To Do lists so that you can cross off accomplishments?
5. Do you feel happy knowing you've helped someone?
6. Are you good at doing several things at once?
7. Do you pay attention to details?
8. Do people often seek your opinion or advice?
9. Are you a connector—keeping in touch with friends or acquaintances by writing or calling?
10. Do you believe in the products you have to sell?

You're a born marketer.

looks too "girlish"—by-passing some otherwise cute items and generally leaving empty handed. When I'm in an expensive store where I feel out of place, personal sales "persuasion" welcomes me in and makes me feel comfortable, as if I (and my credit card) do belong.

By kindly assisting your customer, you can ease them into a sale. Offer help by suggesting different scenarios and how to cook appropriate cuts of meat for the occasion—mother-in-law-coming-to-town, anniversary dinner for two. Paint a picture of the possibilities. And share easy cooking tips.

Put your customers at easy by telling them how properly marbled grassfed meats don't need special cooking methods. De-mystify the meat.

Attention to details. Analyst Arnold Brown believes the best business training a woman can have is motherhood.

"Women bring flexibility, imagination, intuition, cooperation, consensus and a broad, contextual, long-term view to every sector of the business world." Let's add one more trait to Brown's list - attention to detail.

That's why women are good at ordering supplies. We're the ones who post a grocery list on the fridge and make to-do lists.

I'm guessing a woman designed the packaging for Celestial Seasonings tea. How many of you have taken time to read the quips and quotes on the box while waiting for tea to brew? It makes tea time more personal, like a chat between good friends. This subtle detail encourages me to select Celestial Seasonings over other brands that simply use their company names on their packaging.

While men only think of the big picture ("Let's go fishing!"), women take care of the details ("I'll pack a lunch, get sunscreen, buy ice for the cooler.").

Attention to details can create strong bonds between you and your customer.

Belief in your product. You don't have to memorize all the technical aspects of Conjugated Linoleic Acid (CLA), omega-3s, or Trans-vaccinic Acid (TVA) to tout the health benefits of all-grass products. If just pronouncing the names, much less their properties leaves you tongue tied, simply create a poster to display the key benefits, or offer flyers from published articles to make the point. Don't be afraid to get the word out there.

This is your biggest marketing asset. If you believe in the products you are selling and serving to your own family, it's easy to be a champion of grass-finished products. All the healthy benefits that grass-finished meats and milk products have to offer are selling points in themselves.

After all, doesn't everyone want to buy the best for their family's health? There's nothing more convincing than that.

The importance of setting goals

In our busy lives, many people look back after a period of years and wonder where the time went. When they slow down enough to think about it, they realize they regret that they didn't take more time to ... (you fill in the blank).

"Aging experts have determined that regrets in older age increase the risk of depression and a lower quality of life," explained psychologist, Priscilla V. Marotta, founder of the Center of Psychological Effectiveness (COPE).

Attaining all of your goals, she notes, isn't nearly as important as using goals to continually strive for improvements in your life.

Without specific, written goals, people have a tendency to react to events with a quick-fix rather than considering long-term results. That can have disastrous results in time or money spent unwisely.

For example, as a freelance journalist, I was once asked by a national series publisher to write a Bed & Breakfast guidebook on Mississippi. My initial reaction was, this would be easy. I've personally visited many B&Bs, and the travel required would be "in my own backyard." Wouldn't it look nice to see my name on a book about my home state? It would make me "an expert" on that subject, causing other editors to call me for more assignments, resulting in tighter deadlines. (And probably less pay.)

Weighing the long-term options, I realized I would have to set aside what I was doing in order to focus on the B&B project, causing everything else in my life to back up. Expenses (not covered by the publisher) and their one-time-only fee (royalties for on-going sales were not offered) couldn't justify the time and effort required for the fee.

The longer I thought about it the more stressed I felt. Remembering some advice I'd heard years earlier, a sign of success is those things you choose to turn down—I suggested another writer who was hungrier than I was.

That example shows how goal setting can lay a foundation for your lifestyle, and quality of life. Isn't that the goal we all have? It's certainly a starting point for meshing yours with your spouse and family's goals.

Setting goals should be a regular task, not only for you, but for your spouse, and the children who may be involved in the farm operation. Take some time alone to set your individual goals, then set a date to work together on shared goals for your farm and family.

Don't rush. If you're not in the habit of setting goals and writing them down, give yourself a week of "think time." Carry a notebook or slip of paper with you and jot down thoughts as they come to you. Then plan some quiet, meditative time to think about what's really important to you, and list practical steps you can take to reach your goals. Never mind if your notes are damp from the shower or tub if that's the only "quiet" time you have to yourself.

Allan has discovered that the time of day he works on goal setting greatly impacts the result. In the morning, he's fresh and able to list several venturesome ideas. After lunch when he has napping most prominently on his mind, his goals are to quit working, retire. If your husband is like mine, he doesn't like to be pushed into these things, so give him enough time and space to work on his personal goals.

Goals are like road maps. They help you know which direction to take. They can help you monitor how far you've come. If you don't have goals, how can you know you've gotten to where you want to go?

Like a map, there are many crossroads. There are no rules that say you have to stick to the main highway. Sometimes you have to take detours (when something's under repair or you discover the route you're following dead-ends).

Maybe for a lark, occasionally you want to try something different. By switching routes you may discover an unexpected, yet more pleasant, progressive alternative. Just remember, unlike Roman roads and mile markers, goals aren't set in stone. They're guidelines.

When he first started writing down goals, Allan realized after a period of a year or so that those first goals were not his, but had been implanted in his mind by his father or a professor. Once he realized that, he was able to set goals that he wanted to achieve, for himself.

When he sees someone successful, he asks himself, what did it cost to get there? Or what did that person have to give up? Then he asks himself, would he do the same thing if he could? That often makes him realize he doesn't want a particular goal after all.

Goals can set the foundation for your farm and family's

Three little questions:

1. How do you want your life to be?
2. What do you have to produce for it to happen?
3. How does the land need to be to provide the life that you and your family desire?

These are three parts of one's holistic goal. Every decision should move you toward that holistic goal.

Terry Gompert, Extension Education, University of Nebraska-Lincoln Extension, Center, Nebraska.

future. They create clarity of purpose. By working separately and then together, they're a check to be sure you're both "on the same page." And if you're not, there are warning flags indicating areas where compromise is needed. Shared goals can help you work more harmoniously together as a team.

Write it down. Some business leaders believe that writing down goals forces you to think things through, and that after you've gone through that process you can toss the paper. But I don't agree with that. By writing down your goals, you will have a record of not only where you're going, but where you've been. Periodically reviewing your goals will show you how and when your goals have changed, hopefully for the better, so date them.

Be specific. Don't just write down, "I want to earn a comfortable income for my family." How much is "a comfortable living"? What will it take to reach that dollar amount? Maybe it's not a dollar amount but a goal to reach a level of income that will allow you to work half days or three days a week, or take a vacation two or four times a year.

Set timelines. While your ultimate goal may be to make $1 million, set a time to reach that income figure, say within 10 years. Begin with practical short-term steps along the way. "Within one year, we will sell $100,000 of grass-finished products. Our five-year goal will be $500,000."

Step-by-step. List the steps necessary to achieve those goals, both short-term and long. Be realistic. But don't be afraid to dream big. You might surprise yourself at how easily, and quickly you reach your goals, once you've taken the time to figure them out. Short term goals give you a sense of accomplishment and progress.

In setting goals— both personal and business—address the various aspects of your lifestyle. Some category suggestions may overlap. Perhaps you've always wanted to learn another language as a personal enrichment goal, but now, learning another language could help you meet your business marketing goals of communicating better with your ethnic customers.

Or a business goal of creating year-round grazing with a forage chain of grasses could include attendance at a conference on the subject to develop your personal education.

Once you've started listing goals, you may find that you need to break down broad categories even further. For example, Business goals could be subdivided to cover Financial Goals, Production Goals, Management Goals, and Marketing Goals. Personal goals could be separated into Physical Goals, Mental (educational) Goals, and Spiritual Goals.

Prioritize. Once you've set your goals, prioritize, then concentrate on the top three to five.

Plan to revisit your goals on a regular basis—at the end of each year or January when people are most likely to make New Year's resolutions, are easy times to remember.

Dream big. Remember, a rubber band is useless unless it is stretched.

Creating life planning goals

Christian author and concentration camp survivor, Corrie Ten Boom, used to take a small weaving with her on speaking engagements. She would show the audience one side with its tangle of colored threads, but when she turned it around, the other side revealed a beautiful picture of Jesus. The tangled threads represented the confusion of things we can't understand that are happening in our lives. The other side represented God's hand in the midst of our tangled lives.

Life planning is like a multi-dimensional plan for your individual lives and shared business and family goals. Believing that you can get by on love alone is a little like the tangled back side of the weaving. Life planning helps you to know which colored threads go where to create a clear picture of your shared goals.

Don't think you can keep your business and family lives separate, even if you have an off-farm job totally unrelated to agriculture. Life is never so clearly black and white. Here's one example of how life and business overlap—your child gets sick the same day you're having an open house field day. You can't say you'll deal with your child after everyone goes home. You will find a way to care for your child first, even if it means calling a relative to serve as hostess in your place.

Life planning involves overlapping your personal desires and interests with those for your shared business goals.

Dream big. Give yourself permission to dream big dreams, and don't even consider how impossible some of those dreams might be to achieve. Think back to your childhood. What did you want to be when you grew up? Write down everything you always wanted to do but never had the time or opportunity to accomplish. Yet.

Don't settle for average. A little extra effort might take you beyond your wildest expectations.

Pluses and minuses. Write down, then rank, your proudest achievements. Now list and rank your bad habits. You trust your spouse, right? Ask for feedback. Get advice from your best girlfriends. Write positive affirmation notes and stick them on the mirror. Tap into your personal resources. Remember the good things you've done in the past. They can be the foundation of better things to come.

Jackpot. If you suddenly inherited $1 million, how would your life change? How would that windfall affect your business?

Brainstorm. Once you and your spouse have both made your lists and checked it twice, compare notes. Are there things you both want to do? Or do your dreams take totally opposite routes? How can you lend support in some area to help your partner achieve their dream? Do you have similar values regarding money? What would you change about yourself to be a

Setting goals has three parts:

1. Taking time to think about those things that are important to you;

2. Interviewing yourself to figure out how you can attain those goals, and recognizing the degree of commitment you have for them;

3. Then prioritizing those goals so that you can take action steps to accomplish the top three or five on your list.

better person? What would you change in your spouse to improve your relationship? How will you prepare your children to participate in the business?

Worst case scenarios. Make a list of the worst things that can happen, then discuss contingency plans on how to deal with the issues, whether they're drought, disaster, or divorce. Could your partner take over if one of you was disabled? It's important not to dwell on the negative, but planning a positive outcome will deflate the fear associated with the unknown. Having a plan will allow you to take action, quicker, when the unthinkable occurs.

Sarah Young wrote in *Jesus Calling* "Rehearsing your troubles results in experiencing them many times, whereas you are meant to go through them only when they actually occur. Do not multiply your suffering in this way! Instead, come to Me, and relax in My Peace."

Be flexible. Life is full of surprises. Expect setbacks and obstacles, however, realize there are always options. Give yourself permission to change as you and your business grow. If you could start over, what would you do differently?

Image. How do you regard each other? What is your business' reputation in your community? Once you have determined the basic values and goals for the shared vision of your business, turn this into a mission statement. Post this where everyone can see it—on your newsletters, letterheads, signage.

Making your goals three dimensional

Michelangelo was 24 years old when he created the statue of Our Lady and Christ called the Pieta. When asked how he could create such a masterpiece he said he saw the sculpture inside and chipped away all the marble around it. Once his vision was clear the work progressed rapidly.

In setting business goals, you too should take the time to envision success and explore all aspects of what that means to you and then eliminate all the peripheral, mis-directed avenues. Once you have defined success, in detail, then you can work backward to figure out how to achieve it.

The first thing to ask yourself is, where do you want to go? And why?

Let's say you're going to make a dollar figure income your goal. It's not sufficient just to write down an amount or a percentage increase to strive for. You have to consider the entire business goal, return on assets, margins, and where you want to be positioned in the marketplace.

Perhaps you plan to increase your product sales by 10% each month. Factor that out for a year-end result to see if it matches your income goal. By doing some goal searching on paper, you may realize an increase of 15% will bump you up into a position to launch the dreamed-for on-farm store or reach another area of growth. Perhaps it will allow you to quit that off-farm job.

Terri and Bob Blanchard of Old Creek Ranch near San Luis Obispo, California, had a marketing goal to sell all of their farm's production. To meet that goal they realized they needed to make a weekly 400-mile roundtrip to sell at San Francisco Bay area farmers' markets with a demographic clientele to add to their other marketing venues.

The next stage is to figure out the necessary steps to take to attain your desired level of success. What will you have to do to get there?

Here are some things to consider for reaching your goals:

Do you need to sell at a farmers' market or two or more?

Will Internet sales help you achieve your goals?

Can a metro buying club work for you?

Should you invest in a seminar on marketing? Things you learn in schools, conferences and workshops can give you an educational edge over your competition.

In *The End of Marketing as We Know It*, Sergio Zyman wrote that having an aspirational goal is great, but recognize it's only hope, not a plan.

"Marketing is about spending money on activities that enhance the value of your product, brand or service," Zyman stated, "and give consumers more reasons to buy more of it, more often."

Time frame. Play mind games with yourself in setting goals. What is your long range vision for ten years? Where will you be in five years? What do you see yourself doing in one year? Work back from there to decide what needs to be done to reach each of those time markers.

Treat it like a game, but realize you're only competing with yourself, not your competition. If you stay focused—and true—to your goals, you will succeed in the marketplace regardless of what the competition is doing. It's up to you to get better at what you do, add more variety to your sales tactics, and do things that work for you. If you're constantly looking over your shoulder at the other guy, you can easily become tempted to follow their road rather than your own. Whose dream is this anyway?

Don't try to be a copy cat, unless you are using someone whose business is masterful and can serve as a benchmark.

Ari Weinzweig, co-owner of Zingerman's Deli in Ann Arbor, Michigan, had this to say:

"You have to allow yourself the freedom to dream in the long term. For us that means at least 15 years out. You don't want to be constrained by a feeling of having to know how to do it. Most people won't allow themselves a great vision because they are afraid they won't be able to get there. The whole point is not to worry about the 'how' of accomplishing the vision. Stay concentrated on the 'what.' And you have to write it down. It doesn't become real until you write it down."

Ari said they then took their 15 year vision and broke it down into three year segments. These mini-segments had a strategic written plan of how to accomplish the short-term goals.

"A strategic plan is what makes the vision come into being. Without a realistic plan a vision is just a fantasy."

He said building a business was like sailing a boat from Hawaii to California. "You may get blown off course by the economy or some other unforeseen circumstance, but your final destination never changes."

Ari said the most important benefit of a long-term vision is that it tells you what you will not do.

Goal setting tips

Be specific.
Give your goals a time target.
Make your goals measurable.
Sit down and write out your goals now.

"A written vision will keep you from chasing unrelated opportunities. It's very important that you develop a *not do* list to stay focused."

Time out. Allow yourself time to dream, and dream big. Take time to step back, review what you have accomplished so far, then recognize how where you are currently positioned can launch you to the next level.

Do you have a quality product you're proud to sell? Is quality more important to you than quantity? (Not a bad goal!)

Well rounded goals. At the *Stockman Grass Farmer* we have always found that those times when we put money as our only goal, projects are far less successful. When we put our customer's needs first, success follows. The lesson is that money can't be your only goal.

Who has a stake in whether you fail or succeed? Factor this into your goal setting. Foremost are your customers who deserve and should expect the highest quality of goods and service you can present. What can you do to improve things for them?

What about your employees (this could be your children as well as workers not related to you)? Do you strive to give them a pleasant work environment, fair compensation, benefits? How can you make things better for them?

What about the owners—you and your spouse or other partners? Will your goals provide a good return on investment?

Do your goals reflect good citizenship and active involvement in the community?

Are your goals good for the environment due to sustainable practices?

That's a lot of "food for thought" at which to chip away in order to one day have a fully developed, well-rounded masterpiece.

Finally. Write it down. This will not only help you stay on course by using what you've written as a point of reference, it can also help you see the progress you make along the way.

What are you selling?

Quality is job one

What motivates people to buy?

In a marketing survey reported by Jay Conrad Levinson, author of the *Guerrilla Marketing* books, quality came in second to confidence in the seller. "If you offer poor quality," he writes, "marketing will hasten the day that your company closes its doors because people will learn of that lack of quality sooner than otherwise."

How do you create quality?

Product quality. If you don't have a quality product to begin with, you have signed your company's death warrant. Some producers who are "the only" grassfed producers in their market area slip into that comfortable space of thinking their products are "good enough." That may work for awhile, but sooner or later, that "good enough" attitude will catch up with them. Your customers may be loyal to you, but when they find something better loyalty thins like flour-and-water paste.

Quality producers have high standards. They go the extra mile to make sure their grassfed products can stand next to the best in the country. They care about customer loyalty and do everything they can to keep their customers' trust.

Packaging. Never assume that people will equate quality with pretty packaging. Clear packaging will win over the generic white butcher wrap. Customers like to see the product they're buying. Think in terms of environmental transparency and responsibility. Don't use excess packaging that adds to our nation's reputation as a throw-away society. Use re-usable cloth bags. Recycle. And let your customers know things you do to be environmentally conscientious.

Public service announcements. Let your customers know what you do to ensure they receive quality products: USDA inspected; no hormones, no grain, humane animal treatment, in-store-kitchen health department inspections. Show them pictures of your healthy pastures to remind them that we are what our animals eat.

Longevity. Use the number of years you've been in business in your marketing to show them you're here for today and tomorrow, not just passing through.

Education. If you attend a seminar, workshop or any kind of class that helps you improve your business or product, let your customers know about it. Maybe the school didn't hand out certificates that could be framed for your walls, however, you can frame a photo of you with the presenter and write a little caption for the photo. This has a similar impact as a newspaper clipping does. Share your excitement over new things so that your customers come to you as their leader in the grassfed industry. Be the first to tell them what's happening nation-wide. Keep a book of grassfed news clippings. Keep your customers informed about your continual quest for knowledge to improve your business for the sake of your customers.

"Marketing...is what moves goods," Levinson writes. "This may be true, but it moves goods only once. High quality leads to repeat and referral sales. And these are the wellsprings of profits."

How to judge grassfed quality

1. Prepare it for yourself.
2. Sell it to your family and church friends.
3. Then sell it to your friends' friends.

Remarkable products

One of our favorite business books is Seth Godin's *The Purple Cow.*

The concept of Godin's book is that you have to have a remarkable product in order to be noticed. The example he gives is a purple cow. If you ever saw one standing in a field surrounded by all the ordinary black, white and brown cows, you'd slam on the brakes and do a double take.

He says you won't have much success with a product that appeals to everyone, because that doesn't appeal to anyone specifically. "The everybody products are all taken," he says.

For real growth in your business you have to have products that fit a niche and will always be "too" something—products that are too complicated to easily duplicate or produce, or too expensive for some people, but just right for others. He feels the narrower or more focused a product starts, the greater its chance of achieving large scale success later on.

"The way you break through to the mainstream is to target a niche instead of a huge market. You want to create a product that is so focused that it overwhelms that small slice of the market that really and truly will respond to what you sell," he explained.

"After it dominates the original niche, if you're good and you're lucky, that product will diffuse. After it dominates the original niche, it will migrate to the masses."

Because no one can be sure of the size of a new market, initial production has to be small and tentative. However, if this product attracts even a small group of enthusiastic consumers it can grow from its own cash flow because it can earn premium prices as long as supply is less than demand.

And, as long as your competition hasn't tried to copy you.

If you are growing animals for truly flavorful, tender, well marbled beef, you know this remarkable product isn't easily produced. You'll leave your competition in the dust because they won't understand, or take the longer time it requires to produce such quality.

Large producers have large overheads and therefore can only target markets with sufficiently large proven volumes of business. These larger producers tend to have huge capital and emotional investments in the status quo and do not want to participate in anything that could unsettle the value of their investments. As a result, they purposely ignore the upstarts and fervently hope they will just go away.

Joel Huesby of Thundering Hooves in Touchet, Washington, compares it to a game of basketball: The big guys may be tall and good at guarding, but the little guy can nimbly dribble and dodge around and under the big guys.

A remarkable product should form the basis of your marketing. A remarkable product creates passionate customers. In the early stages, excited customers will spread the word—for free—to their friends to bring more customers to you.

What do you do when you've saturated the market with your single remarkable product?

That's the time to add something else.

When to add new products. If you're not constantly redefining your products, you won't have growth and you'll soon be dead with the same-old, same-old. For example, that Sunday roast, spiced up with a new recipe or suggested marinade, can be touted as a new menu solver for middle of the week.

However, sometimes repackaging, or re-introducing the same product is still not enough.

It's important to continually challenge yourself with new ideas and plans, because when the competition wakes up and notices what you're doing, they'll start acting like you. Always be looking for the next new thing.

In *The End of Marketing as We Know It*, Sergio Zyman recommends only adding something new when you're sure your core product/s can't be stretched to meet the needs and desires of more customers.

When you're constantly changing what you're doing, you're more likely to leave the competition scratching their heads in confusion over what you're doing. Think leadership. Think cutting edge. Think of being the first.

Related items. Only introduce new products if they can add to your core product. For example, artisanal bread matches well with aged, raw milk cheeses; selling meat and motorcycles doesn't. It's always easier and more profitable to sell more products to existing customers than to find new customers.

Industrial versus artisanal markets

- Industrial agriculture embraces control and the conquering of nature.
- Artisan agriculture embraces the wildness and unpredictability of nature.
- Industrial agriculture embraces cool-running technology.
- Artisan agriculture embraces heat and human sweat.
- Industrial agriculture seeks out ever bigger markets.
- Artisan agriculture seeks to satisfy the passionate minority.

Allan Nation, Stockman Grass Farmer

Acting on this principle, Fox Fire Farms in Ignacio, Colorado, have added salmon to their lamb and beef line-up. Does that sound a little, well, "fishy"? Not really since their source, Al's Alaskan Seafood has fairly exclusive fishing rights to catch wild salmon on the Copper River. Being wild salmon guarantees the fish were not farm raised. The wild, Copper River salmon complement their all grass organic lamb, organic beef, organic goat, organic free-range eggs and organic wine.

Complementary versus competitive. Be sure the "new" does something distinctly different from your existing product. When you have something new, analyze whether buying "this" will cause your customers to stop buying "that." On the other hand, complementary products can encourage piggyback buys.

"Remember," Zyman warns, "you're not in the volume business, you're in the profit business."

Oops. Occasionally you'll come up with something new that just doesn't work in the long run. Joel Salatin had a group of people in the Washington, D.C., area who were interested in buying a heritage breed of chickens called Barred Rock Cockerels. Joel called them Marco Pollos (Marco after the navigator Marco Polo; Pollos (pronounced *poyo*), Spanish for chicken). However, the time and effort in producing them (keeping a separate inventory, the birds separate in the brooder and field) far outweighed the profit and demand. Wisely he dropped them from his product offerings.

It never hurts to experiment with something new, just keep it small and the monetary investment low until the response shows it can be a proven, remarkable success.

Providing value with values

How many times have you scrimped on a purchase to save money and been disappointed with the result? It broke way too soon, didn't live up to product claims, made you wish you could get your money back.

You get what you pay for.

Savvy shoppers aren't the same as bargain hunters. They care about getting good value for their money. That's not cheap, but it's also not necessarily pricey.

"Selling is convincing people of the value of your product and persuading them to see how it will enhance their lives," wrote Nicki Joy with Susan Kane-Benson in *Selling is a Woman's Game, 15 powerful reasons why women can outsell men.*

Who wants to scrimp when it comes to their own or their family's health? As a marketer of grassfed products, that's the primary selling point to tip the sale between paying somewhat more for a quality, healthy product and settling for the ninety-nine-cent-a-pound greasy, grainfed bargain.

How do you justify the higher cost to a customer who's undecided?

By providing value with your product.

Trust. It starts by building trust and rapport. As a mother, having healthy children is your proudest achievement. You and your family must walk the walk of the talk you talk.

Leon Gorman, the grandson of L.L. Bean was quoted in the biography of the company as saying, "I believe our character, our values, were what separated L.L. Bean from all the others, what created the mystique that had grown up around the company." It was important for the company to have an understanding with their customers, which built respect. The outdoors was the vehicle that allowed their customers to get in touch with their spiritual side.

Be transparent. To gain their trust, strangers want to know about your ethics, how you treat your animals, what the farm's environmental record is.

Don't assume that just because you say your products are all natural or organic that the average consumer will truly know what that means. If you're grassfed, that means 100%. Those in the know will challenge you on this point.

Be proud to give full disclosure—no chemical use, no GMO seeds, no pesticides, antibiotics, artificial fertilizers, no animal products in the animals' feed, no grain. Ever.

Let your customers know your farm is open for tours, either for a scheduled open house, or an impromptu visit. Invite them to come see for themselves, and bring a friend.

State your mission. Do you have a mission statement? You should. Sit down with your spouse and your family and distill your personal values and that of the farm into a single statement. How do you want your company to be remembered? Why are you in business? Your mission statement can come from the answers to these questions.

Having a clearly defined mission statement can keep you on track. When important decisions need to be made, a mission statement helps you stay true to your values. If whatever you're thinking of doing doesn't fit your mission statement, maybe you shouldn't do it.

Use your mission statement every chance you get. Here are a few examples:

■ By nature...With nature...The way nature intended. Open Space Meats, Newman, California.

■ Our mission is to improve the quality of life of our customers by providing superior-quality, healthy food products. Jako, Inc., Hutchinson, Kansas.
■ Raising our children and livestock to the Glory of God. Copeland Family, Siskiyou County, California.
■ Our mission is to create a sustainable farm that produces health and enjoyment for our livestock, our community, our customers, and ourselves. Wolf Creek Farm, near Charlottesville, Virginia.
■ Changing lives, one bite at a time. Hendricks Farms & Dairy, Telford, Pennsylvania.

Is there a particular cause that you believe in and contribute to? Let your customer participate by knowing that a percentage of your sales are diverted to that cause.

I noticed that the logo of my Dawn dish washing liquid has a duck in flight, a swimming seal underneath and says "Rescuing wildlife for over 25 years." There's a website on a smaller sticker that reads, "Help Dawn Save Wildlife." Sure, I'll buy that brand again. (By the way, the label is also multilingual with key statements repeated in Spanish.)

Toot your horn. Testimonials from satisfied customers can inspire confidence in complete strangers. Leave open an album of praise for your products for others to read. Use banner quotes as decoration in your farmers' market booth. Include customer comments in your farm's newsletter. Let the words of others help you sell.

Courtesy and charm cost nothing. Service is a priceless value add-on. Women, especially, are pushovers when it comes to service, particularly when it's unexpected. Any business that isn't service minded will be less likely to receive repeat customers. When a woman's scorned, she not only boycotts that business, but she tells her girlfriends to steer clear of it also. But when she's surprised and pleased, her friends will hear about that too.

If you're selling from your farm, give your younger children the task of carrying purchases to your customer's cars.

On Charles and Laura Ritch's Goose Pond Farm in Hartselle, Alabama, their daughters Abigail and Mary Sidney bagged purchases and sang Happy Trails to You to departing customers. They earned good tips too. (This would probably also work at the farmers' market.)

Cleanliness and special care. At the market in Orvieto, Italy, a farmer selling asparagus ties the spears together with white string and slips a pink rose into each bundle. In that same market, a woman selling eggs and vegetables wraps each egg a customer selects in newspaper and twists the ends as if it were a party favor. The same woman will take a melon, wrap it first in brown paper, then a layer of newspapers, pleating the ends as if it were origami.

Individually wrapping eggs probably won't work for North American customers because we tend to purchase in bulk. In much of Europe food purchases are made daily with only that evening's meal in mind. However, the extra care draws shoppers to that woman rather than others selling the same products in the same market.

Why not tuck or tape a sprig of rosemary onto your white butcher-paper wrapped lamb before handing it to a customer?

How about printing out poems, prayers, riddles, jokes, Bible verse blessings, rolling them up, sealing with a sticker,

What if?

- What if you sold things the way your competition does?
- What if you could charge ten times what you do?
- What if you had to charge ten times less?
- If you were on a national talk show, what would the host say about you?

The Big Moo, Stop Trying to Be Perfect and Start Being Remarkable, edited by Seth Godin

and including them with a wedge of cheese as you wrap it for a customer? Change it up so that they'll be inclined to return for a different message.

And what could be more charming than a drawing or handmade valentine your youngsters have created for your customers? Have them available in hand addressed envelopes, "For You." It's another subtle statement that yours is a family business. Women particularly like to know their purchases are supporting the life of a farm family.

Character reference. Discussing L.L. Bean, Leon Gorman believed, "If we could keep adding value to people's lives, we had an obligation to grow." Who you are should be expressed through your personal style of doing business. Make it a point to treat your customers the way you want to be treated. This alone can be your most important asset and competitive advantage.

Authenticity really matters

Isn't summer great, when all those fresh-picked fruits and vegetables are available?

We always buy way more than we should at farm stands and farmers' markets. Then returning home wonder how to find a workspace in the kitchen when all the bags of fresh corn, tomatoes, cucumbers, eggplant, and blueberries crowd the countertops. Miraculously, within days it's all consumed and we're ready to buy more.

We wanted to take some peaches to a Fourth of July celebration, so we stopped by a fruit and vegetable stand near our house in the community of Oak Grove. The peaches were sweet and juicy. Under the watchful eye of her grandfather, the young girl gave us a selection of ripe and nearly ripe fruit to last through the week.

"Why don't you sell at the farmers' market?" Allan asked.

"Oh, we're not farmers," the grandfather replied. "We live right here in Oak Grove and only buy things we can resell."

Not farmers? Allan and I looked at each other, hiding our disappointment from the vendors. Reselling from a distributor, made them only slightly better than a roadside Wal-Mart. We paid for our purchases and drove away, disappointed.

Joel Salatin is a great role model for authenticity. Not only have other farmers adopted his Salatin-style rolling poultry pens, but he has a customer base ready to rally around him because of what he stands for. "They are buying more than chicken and eggs from him," said Mike Lorenz, an abattoir owner in Minnesota. "They are buying into his belief system."

Authenticity. It's priceless.

I have no doubt the farmstand family weren't trying to be deceptive. After all, they openly told us they weren't farmers. And there's nothing wrong with selling things that don't come from your farm. It's important to "load the wagon" with a variety of products to satisfy your customers and give them as much variety as possible to keep them coming back and buying as many things from you as possible. Even L.L. Bean once said, "You can't do business with an empty wagon."

However, if you're going to sell things from other farmers or vendors, let your customers know without making them ask. This family could have made signs saying, "Chilton County, Alabama, Peaches," "Lucedale, Mississippi, Watermelons," or simply, "Florida Corn."

Recognize that farmers' market newcomers are somewhat skeptical initially. The more you can tell them about your product, the easier you make it for them to buy. For example, at our farmers' market there's a woman we call the "goat lady." She sells cheese, milk and soap. But we don't know if her goats are purely browsers or if she feeds them grain.

Authenticity

"Axiom 1. If you *are* authentic, then you don't have to *say* you're authentic.

"Axiom 2. If you *say* you're authentic, then you'd better *be* authentic.

"Axiom 3. It's easier to *be* authentic if you don't *say* you're authentic."

From *Authenticity, What Consumers* Really *Want* by James H. Gilmore and B. Joseph Pine II.

A few stalls away a family sells flowers and eggs from ducks and hens. They have flyer handouts and a notebook with photos that show their free-ranging hens and ducks pecking through the grass.

Authenticity. It's one of a kind.

How many people made a point of visiting the Bass Pro Shop when there was only one in Springfield, Missouri,—called the Grand Daddy? Now they're everywhere, including one in Canada. Customers still patronize their stores, but they no longer go out of their way or plan their vacations with a side trip to visit the one and only.

In an effort to capture more market share, many companies clone themselves. Then after over-extending, the next thing you hear is they're closing down stores. It happened to K-Mart, Macy's, Nieman Marcus and others. The clones carry the same products and have the same decor, but they're missing that intangible atmosphere that comes with being an original. People don't talk about clones. The buzz is always about things that are unique. When Starbucks announced closure of 600 stores in 2008, their independent rivals outlasted them, because they were different.

Here's an off-the-wall idea to make your one-and-only market stall stand out from the competition. Take a cue from bookstores that place couches and chairs to invite customers to hang out with a book rather than just dash in and out to make a purchase. Set up an umbrella table and chairs on an indoor/outdoor rug (green if you can find one or handmade, soft and loopy). Encourage patrons to take off their shoes, wiggle their toes and stay awhile.

Give them samples as a reason to linger.

You might even want to encourage your potential customers to paint or doodle on the tabletop or cover it with butcher paper to encourage them to personalize the space. Keep a box of crayons handy for big and little kids.

Due to farmers' market rules you might have to buy a second booth space for your inviting seating area.

Authenticity. It's unreal.

How many products on grocery shelves and TV advertize that they are "real"? Anytime the word "real" is used in marketing, chances are it's not. Would your customers buy Egg Beaters, labeled as containing 99% real egg whites? What's in that other 1%? And what happened to the yolk?

Why would you buy "Real (dehydrated) Blueberries" in a boxed cereal when you can toss in berries freshly picked, or make your own cereal using a combination of whole grains?

This has to be an oxymoron—"Real VELVEETA CHEESE" used in Kraft's Velveeta Shells & Cheese. Or how about a box of dry "Real Non-fat Milk." Those who care about their food know that if it didn't come from grassfed cows giving raw, whole milk, it's not really milk or cheese at all.

So, let's get un-real. Let's deliver just plain delicious products. And let your customers know about it! No grain. No distillers' waste. No corn—a big selling point for people with corn allergies. Foxfire Farms sells "No-Soy Eggs" faster than the hens can lay them.

Paint a picture of your farm's sparkling, clean water and vibrant pastures swayed by fresh breezes. This is where your shiny bronze cows nibble tender salads of mixed grasses and sweet clover to bring your customers unadulterated milk that goes into the rich butter, thick cream, aged cheeses or summer ice cream. Doesn't that make you proud to be a grass farming family?

Be the brand

"You are not in the meat business," said Mike Lorenz, owner of an abattoir in Cannon Falls, Minnesota. "You *are* the product."

Urban consumers have their own doctor, lawyer, dentist, Lorenz pointed out. They should have their own farmer too, and that farmer should be you.

Everything you do and produce should reflect who you are - your heritage, the history of your farm, your values, purpose and mission statement. Think in terms of *being* the brand, and look for ways to celebrate what makes you and what you sell stand out from the competition.

If something doesn't fit the core value of what you represent, don't include it. Remember, you might have to make a "Don't Do" list to stay focused.

"The whole reason for creating a brand is to get consumers to identify a number of desirable qualities and traits with your specific product," wrote Sergio Zyman in *The End of Marketing as We Know It.* Of the examples he gives, my favorite is Arm & Hammer Baking Soda. You can use it for baking cakes or biscuits, as an air freshener sprinkled over rugs or with an open box in the fridge. It can be toothpaste. I use it as a cleanser to remove stains from stainless steel and countertops. And it works great to clean drains when splashed with vinegar. What a product—you buy it to *throw it away*!

Here's another example. S.J. Popeil, the father of the Dial-O-Matic, the Chop-O-Matic, and the Veg-O-Matic made a best seller out of the Pocket Fisherman. "It's not for using; it's for giving."

How can you compare that to the uniqueness of grassfed beef? Make the product the star.

Be honest, not cute. How many consumers have bought milk because of the milk-mustache-on-celebrities ads? What does taste and healthfulness have to do with California's giggling Happy Cows (who, by the way, are usually standing in yellow pastures)?

Give your customers reasons why they should buy your beef, lamb, cheese or raw milk. What makes it different, better or special compared with your competitors?

Grassfed beef has flavor. It doesn't taste like A-1 or Heintz 57 steak sauce. It tastes the way granddad's steaks used to taste, which evokes nostalgia and wholesomeness. It has no grain, no preservatives, no antibiotics. A good grassfed steak reminds me of the T-bone chops my grandmother used to pan fry when I visited. She bought them from the butcher in a little grocery store a few blocks from her house. In the 1950s, there was another little grocery store a couple of blocks from her home in the opposite direction. She bought things fresh, just before she planned to use them. It's the way many Europeans shop—daily, at open-air market stalls, preparing what's fresh and available that day, which alone evokes memories of the Old Country for some immigrants.

Good enough is not good enough. Your marketing should educate your customers that the average grocery store today has lousy meat and cheese, in spite of their slick hyper marketing. Your customers should expect better for their families. In today's marketplace, consumers have opinions, so you have to tell them how to choose.

"Sameness doesn't sell," noted Zyman. "The value of your product will be determined by its differentiation from the competition in ways that are relevant to consumers."

Ask yourself what aspects of your history could be celebrated? Make a big deal out of it. While I'm not advocating corn for our all-grass livestock, I found it interesting to learn that The Corn Palace Convention & Visitors Bureau in Mitchell, South Dakota, decorates the entire town hall annually with corn to celebrate the 1000-year-old history of corn growing in the region.

Your future success depends on understanding the unique character of your brand and expressing it through the farm's activities, both on the farm and in the community. Are all of your products true to the niche your brand fills?

Looking good. Your personal image and that of your marketing venue should reflect a clean, attractive, inviting image for your customers. Make sure your customer's first experience is a positive one. The words on your labels and the materials your products are packaged in should be clean and drip proof—clear plastic or white butcher paper. Food safety and cleanliness is a huge issue.

Sounding good. While you're focusing on looking good, make an effort to sound good too. You should know your products inside and out and be able to answer any question that comes up. Be aware that your customer may not be the only one listening to your conversation. Practice patience and courtesy.

How to create brands of distinction from these combinations:

1. Customer self-image reinforcement.
2. Customer values alignment.
3. Customer segment focus.
4. Unique product/service purchase or usage experience.
5. Outstanding customer service.
6. Best overall value.

Brandingstrategyinsider.com

Emotional branding. When a woman is in the "deciding mode" before reaching the "buying mode," several things can unconsciously affect her buying decision. Ask yourself these questions:

Does your brand evoke powerful, positive memories? Is the aroma of grilling samples reminiscent of fun, family cookouts? There's a reason companies give their products "good memory" names. Take shampoo as an example—there are Ocean Breeze, Summer Rain, and Honeysuckle choices.

Remember that every woman has, as Mary Lou Quinlan termed in *Just Ask a Woman*, a personal Board of Directors. These are people she confides in and whose opinion she values—close friends, family, parents, or someone in the food services business. For women, "It's not just business, it's personal," said Quinlan.

Logos. Do you need a logo? Well, no. However, if you plan to be in business a long time, a logo can build brand identity. For Green Park Press, our book publishing division, we created a green, oval logo with cows grazing in a green park mountain meadow. It becomes more recognizable with each book we publish. It's our symbol of environmental awareness and our commitment to green pasture grazing. We use it in our books and on our public relations materials.

Jingle jangle. You know that song you can't get out of your head? Hire a songwriter (students love these projects) to create a jingle for your farm. Or simply use a phrase that can separate you from the competition and by constant repetition (on your order slips, invoices, farm signage, banners, truck decal) will be remembered. For their company, my friend's grandchildren pop up at the end of their ads (TV and radio) to say, "That's Daniell with two "Ls"."

Include your mission or vision statement with every marketing item. Remembering your mission helps customers remember you.

Be a legend. Have any celebrities or chefs bought your products? Have you or your farm been written up in the press?

Laminate the article. Blow it up. Make a poster out of it. Or put clippings in a book that customers can flip through.

At the Dupont Circle Farmers' Market in Washington, D.C., Joel Salatin's marketer used to put a stand alone placard in front of the booth that read, "As featured in *Smithsonian.*" Make an effort to spread your good news.

Create your own buzz if you don't yet have any clippings. Set up a taste test with local chefs, or man-on-the-street samplings and evaluations. However, make sure you have an exceptional product before you try this or that first impression your customers have could backfire.

Novelty items sell simply because they're different. In a recent year at our church's Irish-Italian Fest, someone thought of selling meatballs in ice cream cones. Everyone had to try one, and the booth quickly sold out. Give people a reason for talking about you.

Make sure you're remembered. Leave a trail of flyers or postcards everywhere you go—in taxis, buses, airplanes, the library, the post office, doctor's office, restaurants, and church.

Have business cards printed to reflect your brand. Darcy Rezac, author of *Work the Pond,* carried aluminum business cards when he worked for an aluminum company. Consider a die-cut shape, more expensive, but it will make your card stand out and be remembered.

Cards can be created on the computer in a pinch, but when they're professionally duplicated by an office supply company you'll be more likely to pass them out freely.

Use both sides of the card to get your message across.

Make cards for everyone, including (especially) the children if they help you in your farm marketing.

If someone hands you their card, take a cue from the Japanese and spend a moment studying it, noting the colors, design, or anything that will help you remember the person's name.

Put your cards in laminated stands, or a little basket, anywhere that keeps them visible and easy for customers to pluck into their purse or pocket.

Include one with every product or purchase. Tape it to the wrapping of packages.

The old way of doing business would be to follow up with a phone call whenever you collected someone's business card. The new way to network reflects a genuine interest in the other person. By finding out if there's something you can do for them, expecting nothing in return, you create a genuine interest in the other person and a more binding connection.

One direct marketer quilted hot pads in the shape of a barn and had her farm's name and phone number on them. She wanted her customers to always have them in their kitchens as a reminder.

Keep things fresh. Don't be quick to jump into a new market, with a second location for example, just because it's possible. It's better due to brand loyalty to keep working your existing customer base.

Keep a steady stream of new ideas in front of your customers to give them ample reasons to buy your brand. Ask yourself, "What if we were just entering this market?" Should you change your packaging? Let your customers know about it. Different things will appeal to different people.

Mart Martin, creative director for the Jackson Spalding Agency in Atlanta, and a former marketer for Coca-Cola™ said, "If you don't brand, you will be branded (by your customers)." Just be sure your brand is the one *you* want it to be.

Selling benefits

Competition gives customers choices. If you don't help them with their decision making—by giving them more information—they're going to select by the one thing they understand—price.

Your products may not be that different from your competitor's, however, your role is to make the customer *think* yours are better, or more special.

Did you ever consider the fact that you're not selling meat, or milk, or cheese? You are being of service to your customers.

They want to know, "What's in it for me?"

To answer that, give them lots of choices. Sell them benefits.

Here are grassfed products biggest benefits:

It's good for you. "We are gardeners caring for the King's garden," wrote St. Francis de Sales. The most wonderful feeling in the world is knowing that you are promoting good health by selling grassfed products.

Let your customers know that choosing grassfed over grainfed can help prevent heart disease, cancer, and diabetes.

Are your customers weight conscious? They can quit counting carbs and calories since grassfed products will help *maintain* weight, naturally. And the healthiest attributes—CLA, vitamins A, E, omega-3s, and more—are in the tasty fat of whole milk, cream, butterfat, and juicy marbled meats.

Research has shown that grassfed cheese and milk pack the most healthy attributes. A daily serving of grassfed meat and milk or cheese can meet the minimum dietary requirement for CLA.

Always stress that your products come from animals grazing 100% pasture. Tell them over and over or with signage that these products are free of growth hormones and antibiotics. Let them know that any animal that requires treatment is pulled from your grassfed program.

How about this from Randy Sipe's banner-wrapped van: "Ingredients: Grass, Water, Sunshine."

Absolute traceability. With all the food recalls in the news, consumers care more than ever about where their food comes from. Make it clear that you are selling products you can trace back to a single animal. Be sure they realize you know

Here are some reasons to celebrate the benefits of grassfed:

- By being local, saves fuel and transportation costs.
- Grassfed meats have a distinctive flavor.
- It's versatile. It can be used in simple recipes (ground meat for sloppy joes) or for sophisticated occasions (tenderloin for two).
- During the sports seasons, do your guys watch football, play football, dream football? Then they should eat grassfed while tailgating. Cook a pot of grassfed chili for the big game.
- It's beyond organic.
- It can turn vegetarians (most don't eat meat for health reasons) into meat lovers.
- It's trendsetting. Grassfed is today's buzzword, one which we predict is here to stay.

each animal's history. If you are selling products from like-minded grass farmers, let your customers know through labeling that you can trace a particular cut back to a single animal.

Buy local. Explain why local is better than industrial agriculture. It benefits the farmer, who the customer can meet face to face—no corporate hideouts here. It builds relationships and trust. By supporting the farmer, customers are protecting farmland from being devoured by housing projects or urban expansion. It saves fuel that would otherwise be spent hauling products halfway across the country. It's good for the environment. And, it keeps money in the community.

Easy to use. Dispell the myth that grassfed meats require longer or different cooking methods than customers may be used to with grainfed meat. What's good on the grill will work just as easily and have more flavor using grassfed meats. Assure them that grassfed meats work well in all of their favorite recipes.

Lean meats are another story. They require slower, moister cooking because they generally come from animals that have been harvested too early, before they could be properly marbled.

One on one. Let customers know that if they call or email you someone will have an answer for their questions or confirmation of their order within a reasonable time period. If handling phone calls is taking too much time away from your work in the field (what a nice problem to have!), follow Joel Salatin's example. He hired someone living off the farm just to deal with phone orders and routine questions. This can easily be done using Call Forwarding.

Shortcuts. Anything you can do to make things easier for your customer will help boost sales. For example, many people (me!) don't know how to properly cut up a whole chicken. Even though it may be less expensive to buy the whole chicken, this lack of expertise holds them back. Find an illustration—or make your own with photographs—of the proper way to do this and include these with purchases.

Or include recipes intended for roasting or grilling a whole chicken. Always try to do the thinking for your customers. The more you can do for them, the less decisions they have to make before buying.

Time savers. Can you think of ways to shorten kitchen time? Is there anything you can sell marinated, pre-cooked, or ready to serve? How about skewered kebabs ready to grill? Or packages of frozen meatballs?

George and Eiko Vojkovich of Skagit River Ranch in Sedro Wolley, Washington, sell pre-formed hamburgers by the pound with a 50-pound minimum order.

Some producers sell individual chicken parts because they know their customers don't want to take the time to cut up a chicken themselves.

Neil Hamilton, author of *The Legal Guide for Direct Farm Marketing* points out that as you carry more products (processed and non-food items), particularly those produced by other farmers, you will be considered a "store," subject to an additional set of legal issues. Check things out with local, state and USDA regulations first.

Menu planners. The more variety of products you have to sell—lamb, beef, chicken, pork—the greater the possibilities for menu planning. Since most farmers' market shoppers are searching for something to serve that evening, take the guesswork out of it for them.

A Featured Recipe of the Day could coincide with an over supply of a particular cut of meat, or hamburger. Change it up to encourage shoppers to check out what you have each time. Suggest times when serving it would be right—New Year's Eve, Valentine's Day, ethnic feasts.

Offer free recipes to go with various cuts of meat.

Budget minded. Offer smaller portions and price variety based on the cut of meat. To begin with you may stress that no minimum purchase is required.

Put together Sampler packages of various cuts, enough for a week's worth of menus.

Although farmers' markets are not the best places to find customers who want to buy a quarter or more of a whole animal, repeat customers may be encouraged to learn that's the most economical way to purchase meat. Ideally you would like to sell a customer a two or three-month supply each time they buy from you.

Product support. If you're like me, you prefer to buy a computer from a company that offers free after-sale support. Sounds impressive, doesn't it? Although you may never have anyone ask for it, let your customers know you're a phone call or email away if they need "support" on how to prepare one of your grassfed recipes.

Offer a guarantee. Let customers know that if they're not satisfied you'll give them their money back. A personal guarantee shows the confidence you have in your products, and makes the consumer more likely to put their trust in you.

If you make the extra effort Ellen and James Girt at River Run Farm in Clakskanie, Oregon, do, you'll never have to return anyone's money. Ellen personally tests a ribeye from every carcass for tenderness and taste. By the way, their grassfed beef won a *New York Times* taste test with a wide range of points between their steak and those that came in second and third.

Every morning when you wake up, think of new reasons why people would benefit from buying grassfed products. Imagine the possibilities. It's not just for the big three—breakfast, lunch and dinner. It's for snacks (beef tacos, wraps), picnics, bridal and baby showers. And more.

Naming names

Many of you may already be wedded to your farm name. Perhaps it has been used for multi-generations. However, if you're new to grass farming and wondering what to name your farm or products, here are some suggestions.

Easy to find. An ad in a major national newspaper showed a well dressed couple and these words, "Brioni. Be one of a kind." Mens' wear? Women's fashions? Jewelry? In tiny print was a website, which my computer blocked for security reasons. A Google search turned up men's Italian clothing. Who knew?

While it's tempting to use your own name for your farm, it doesn't give buyers a clue as to the type of business you're in. Choose a name that reflects what you do. Make it easy for customers to find you by the name you use. Grassfed Meats by (Your Last Name) Family Farms is 100 percent identifiable instead of putting your name first. Someone who sees just your name and doesn't know you won't remember your name when they're looking for you in the Yellow Pages. They will remember the product category you sell. Here are some random good business name examples from the phone book: Cookie Classics, Pastry Garden, The Cabinet Studio, Tranquility Day Spa, U-Haul.

Can anyone guess what these businesses offer? Nordan Smith (natural gas), Baucom Stacey & Greg (gift shop),

Earthtone Designs (swimming pools), Zip Express Installation (TV).

This name is just plain ugly—Cajun Kitchen & Drywall. Yes, the business categories are clear, but who wants to buy from someone selling such diverse and unrelated products?

Be clear. Be obvious. Don't make prospects guess.

Easy to pronounce. We purchase our cheese from the Yegerlehner family in Indiana. I have trouble spelling their name as much as pronouncing it. They have wisely named their grassfed cheese business The Swiss Connection.

Whatever you choose, be sure it doesn't sound confusing when you say it out loud. Several years ago a friend of ours was involved in the promotion of a Graze Fest in a market that largely knew nothing about the grassfed industry. Using her contacts, she booked several radio ads. Most people thought the ads were saying it was a "Grays Fest." Was it a meeting of Confederate Civil War Reenactors? Were a bunch of seniors getting together for hair color makeovers? The confusing name hurt the ad campaign.

Easy to grow into. If you're thinking of adding multiple enterprises to your farm business, give it a name that will allow for eventual growth. When you become recognized as a beef supplier, customers may not realize that you also sell lamb or cheese or chickens. Joel Salatin's Polyface Farms covers all the bases.

Easy to remember. Have you built a reputation in your community? Remember the example of the Daniell family from whom we buy our pre-owned cars. Their ads always end with, "That's Daniell with two 'Ls'". Do you have name recognition for people to remember you?

Authenticity. Names should engage memories and be used to recall an earlier era, simpler times, evoke nostalgia. They can also reflect the founder. Don't you think of fun when someone mentions the Walt Disney Company?

Naming products. Joel Salatin enjoys naming various Polyface products and farm tools. These pun-derful names not only make people smile, but help people remember them. This is one way he never has to worry about copycat vendors.

His moveable sheep shelter was a Ewe-go or maybe it was a Lamborghini. His mobile shade is a Shademobile. He has been so successful with his moveable pens for chickens that everyone who copies his design calls their pens Salatin-style, and he has no beef with that. His grassfed pork is Pigerator Pork.

To eliminate customers' questions of "how many pounds is that?" he assembled boxed products. Each box has a quarter of meat in it, and he admits he had fun coming up with the names: All America—a "Butcher's Choice" type of arrangement with bones, organs, steaks and roasts; Slow Cooker's Dream—contains stew, chuck and boney pieces suitable for stewing; Country Club—no organs or bones, more ground than Slow Cooker, more boneless meats and some steaks; Deluxe Delight—high end boneless meats, no organs; Meatloaf Mania—all ground; Chop Lickin'—pork chops, Boston Butts, sausage and spareribs; Loin Party—boneless

Additional naming guidelines

1. Put a positive spin on it. Be sure there's nothing negative about the name.
2. Does it sound like a name that's already familiar?
3. Avoid trendy names unless you want your business to be dated.
4. Using your own name helps keep the competition at bay.
5. If you plan to use the Yellow Pages in your marketing, choose a name that will put you at the head of the line such as AAA Automobile Club.

Guerrilla Marketing Weapons, Jay Conrad Levinson

tenderloin plus everything but pork chops in the previous box; Sauce Pan—sausage, ham roasts, less Boston butt and a bone-in tenderloin. Assemble your own prepackaged items based on your customer sales and items hard to move, and dream up your own clever names.

Place names. Just as true champagne only comes from Champagne, France, place names can grow into brand names: Land O' Lakes butter is one example.

Livestock. When giving farm tours, the number one question children ask Alabama producer, Frank Randall is, "What are the names of the lambs?" His reply, "#7234." Naming every animal on the farm is much more personal if the numbers are low. The names can be educational. One farm family used naming the animals to teach their children the names of states. Another year they gave the animals names of rivers. Another producer gave each year's calf crop names starting with A, and the following year, names starting with B. That made it easy to track their age.

By the way, United Kingdom researchers Catherine Douglas and Peter Rowlinson of Newcastle University showed that giving dairy cow names and calling them by their correct name increased milk production yield by an average of 258 liters a year.

Be everywhere. Be sure your name and contact information is on every printed item—invoices, recipes, handouts, newsletters, signage, truck decal, re-useable bags. Even if you use plain brown bags you can have logo decals printed so that they can be stuck on to personalize the bags. That way your customer becomes a walking sign as she meanders through the farmers' market. One summer we picked up bread in bags our baker's grandchildren had decorated with crayon pictures and messages about "Mimi."

As James Gilmore and Joseph Pine said in *Authenticity*, "A good name conveys something real and specific about a company, product or service." Be real.

How much should you charge?

Pricing

A farmer went to buy a truck he saw advertised in the newspaper. The salesman filled out the paperwork and handed it to the farmer.

"Whoa " the farmer said. "That's not the price I saw."

The salesman explained the truck was loaded with extras—power steering, brakes and windows, automatic winch, 4-wheel drive.

Since the farmer really needed the truck he signed the papers and drove it home.

Several months later the salesman showed up at the farm."My son needs a cow for a 4-H project," he said. "Do you have any for sale?"

"Yup," the farmer replied. "They start at $500."

The saleman picked out a cow, took out his checkbook and started to write the check for $500."

"Whoa " the farmer said. "That's not the final price. That cow has extras you have to pay for.

"Let's see. There's the Basic cow $500; Two-tone exterior $45; Extra stomach $75; Product storing equipment $60; 4 spigots—$10 each; Leather upholstery $125; Dual horns $45; Automatic fly swatter $38; Fertilizer attachment $185. That'll be $1113 total."

When it comes to pricing you have to look at everything.

What's your cost? The first thing you need to know is the cost of production, field to harvest. That's your base. Bear in mind that retailers work with 35% to 50% margins with grassfed products. Direct marketing allows you to keep those profits for yourself by eliminating the middle man.

According to Mike Lorenz of Lorenz Meats in Cannon Falls, Minnesotta, a 1200-lb grass-finished steer will have about 420 pounds of salable meat. "It requires about four people to eat one 1200-pound steer a year," he said, "so that gives you a target market size you need."

Exceptional or rare products can most easily garner premium prices. But watch your costs to be sure they're not the reason you're pricing higher.

Product numbers. Direct marketers who are starting out should first sell to their friends, their church mates, and their friends' friends, before ramping up production. This is the testing time to be sure you have a remarkable product that your family and friends love. This also builds confidence in what you sell.

Begin by selling a whole carcass—that's one product. Or offer halves—that's two. Selling quarters gives you four products.

Be aware that the average custom processed beeve has 21 different cuts of meat in it. A steer has only 24 ribeyes and 24 New York strips. Small producers who offer individual cuts of meat risk running out of something a customer wants, and as a result lose the sale. Plus, there's all that hamburger left over. One solution is to sell bundles of meat as "Gift," or "Starter" packages.

"Simple systems make money. Complicated ones don't," Mike advised.

Comparison shopping. Next do some comparison shopping at local supermarkets and specialty meat shops. Check out mail order catalogs such as Williams-Sonoma or Schwan's®, which offers home delivery. (They even have a "Farmer's Market" section.)

Find out what your competition is charging, and compare prices from other areas of the country by networking with your fellow grassfed direct marketers.

Identify your customers. Are you in an affluent area, the kind where, as Mike Lorenz says, they're willing to pay big bucks to brag to their friends about how much they paid for ribeyes? (No one brags about what they paid for ground beef.) Simply stated, where do your customers live?

Are your customers health-conscious moms who want the best for their families and will pay a little more than supermarket prices?

Factor in your fellow marketers' customer bases—are they selling in upscale locations, or high traffic areas? Do they offer more than you have? In your local area, what's the highest price of similar products?

Buying habits. People buy for quality of life, pleasure, and to beautify their home. Because of that, items that are not price-sensitive are speciality or new products, add-on accessories, something seldom used (special occasion foods), something to use and replace, for luxury or status, unique products that are not mass-produced, for recreation or hobbies. Put the emphasis on the value you are selling, not the price.

Choices. Customers like to have choices, however there's a delicate balancing act on how many to give them. Too many will overwhelm them, so keep it simple, but offer more than one option. Minnesotan Brad Wedge said any business that is focused on price will reduce the choices they offer.

Decide on your strategy. Joel Salatin considers being lower cost rather than selling high-as-the-market-will-bear pricing will encourage more consumers to buy into the grassfed world. (Thank you, Joel!) His pricing strategy is to build a loyal patron base.

"If we had raised our price to artisanal grass-finished norms (initially) our patron base would be half its size," he wrote in the *Stockman Grass Farmer.* "And we'd be forced to spend much more time marketing instead of enjoying the satisfaction that comes with healing more acres of farmland."

That can work for you too. Just be conscious of what your strategy is, rather than pricing haphazardly, or mindlessly trying to top your competition.

One couple in Missouri, adds an extra $1.00 per chicken for those sold at the farmers' market. This charge compensates them for extra handling and freezer space, and encourages customers to buy direct from the farm.

A survey by the American Meat Institute showed that only 20.7% of respondents eat natural/organic meats exclusively. However, 63% said they would eat more if the price was closer to commodity meats—food for thought.

Start high, move lower. Don't undersell yourself. It's

Hamburger helpers

So, what do you do with all that hamburger?

Andrew Johnson of Milford, Texas, grinds the whole animal into hamburger. That's back to selling one product. But whole-animal ground meat can be tastier because it includes better flavored cuts, a good niche marketing point.

When he was selling at San Francisco's Farmers' Market, Mark Keller cooked the ground beef and sold Organic Beef Hamburgers for $6.00 each.

Morrilton, Arkansas, lamb marketer, Ed Martsolf, said, "The best way to get rid of low value ground meat is to sell it as a high value something else." He molds ground lamb into a round patty and wraps a strip of bacon around it and calls it Lamb Filet.

Find a market for hamburger before you try to sell the rest of the animal.

much harder to raise prices without customer resistance than to realize your high price is too much and needs to come down. Customers will respond to lower prices, especially when you point that out to them.

Another way to approach this is to ask, "What is the customer willing to pay?" and then work backward. Anyone who doesn't value the time and effort required to produce quality products is not the kind of long-term customer you want to keep.

High quality and extra services. Often we are too price sensitive. Brad Wedge said that 68% of customers leave because of poor customer service, 14 % due to poor product quality, and only 9% because of price, yet 37% of businesses fail due to inadequate pricing.

Speaking at a Minnesota Direct Marketing Conference he listed a number of priceless value-added services that keeps customers coming back—respect for their time, free delivery, after-hours service, accepting credit cards, a nice atmosphere, and location.

Raising prices. Perhaps you're selling multiple products—chicken, lamb, beef—and find that the cost of production has risen. Rather than raise everything, move prices up on one item only, as Joel Salatin does. That way if you encounter some customer reluctance to buy, you can point out that the prices of everything else have remained the same.

When it is time to give yourself a raise, follow Joel's example. He uses media news that has reached the mainstream public to tie in to increased prices. When the public starts hearing more and more about ethanol, high corn prices, fuel surcharges, and drought, for example, they want to know if you're affected by it. Being a direct marketer of grassfed products should insulate you from most of those issues, but they can still be good pegs to justify slight price increases.

The adage, you get what you pay for, applies here. If you price them too low, prospects might wonder if there's something wrong with your products. Do you really want bargain hunters, or customers who appreciate quality? When you value your products and put an above average price on them, you give the impression of quality and something special.

Repackaging. If you quickly sell out of a particular cut, it could indicate that it's priced too low. Bump the price up appropriately. By the same token, things that are slow movers need to be priced lower. The goal of a direct marketer is to adjust the prices of various cuts until you sell the whole animal without a surplus of one thing leftover.

Mike Lorenz recommends selling a Beef Box. He said only one percent of the population has the freezer capacity to store a quarter animal, even less who can buy a half or whole. The Beef Box averages 30 lbs, costs about $200 and contains such items as 2 rib eyes, 2 sirloins, 3 roasts, one-pound packages of ground meat and whatever else you need to move. With 24 ribeyes per animal, you can assemble 12 Beef Boxes. Do not include tenderloin. Save this expensive cut for stand-alone sales.

A Pig Box will average 30 pounds, cost $250 and contain such items as 1 rack of ribs (hams and ribs are the limiters on a pig), 4 boneless pork chops, 1 boneless loin roast, 8 chops, 1 pork picnic roast, 1 ham steak, 3 packages of bacon, 2 packages of wild rice, seasoning packages and recipes and so on. Add whatever you need in order to bring the package up to 30 pounds.

Avoid discounts. Shoppers who buy when there's a discount, will wait for the next discount before buying again. Price promotions don't build long-term support. (I wait for Sunday newspaper coupons before I replenish my makeup and buy shoes.)

Having a "Special" is not necessarily the same as a discount. This could be offering something you don't normally have for sale, such as a seasonal or limited item. It might also be a product you want to test market before adding permanently into your product selection.

Don't slip into the habit of cutting prices at the end of

the day just so you won't have to haul unsold items home from the farmers' market. Shoppers will expect it and will wait until the last minute to buy, if they're still in the area.

Make it easy on yourself. Price your products in round numbers to speed sales and eliminate the problems of making change.

Show don't tell. Post your prices so that they can easily be seen by someone who's browsing past your booth or in your farm store. Don't make your customers have to ask how much something is or end up putting something back because they didn't know the cost until they reached the cash register. This is not only a courtesy to them, but will save you time by reducing interruptions about what things cost.

Focus on profit, not volume. "You absolutely cannot spend more money every year just to keep your old volume," writes Sergio Zyman in *The End of Marketing as We Know It.* "The cost of the volume that you already have should go down the longer that you have it." Zyman also warned, "You can move a whole lot of product if you give it away."

The bottom line is price can't stop passion. True believers don't care what it costs.

Responding to price objections

No one enjoys confrontations, but because women typically are peace makers who look for ways to cooperate, and men tend to be competitive, searching for ways to dominate, women are generally best at overcoming objections.

Today's health-conscious super-moms are more likely to pay a little more if they know something is good for their families. When prices are high, the consumer's expectations and trust level are heightened.

Often, objections are simply a sign that the customer needs more information or justification for the purchase.

Objections should be music to a salesperson's ears, as involved people raise objections, writes Nicki Joy with Susan Kane-Benson in *Selling is a Woman's Game*.

The key in selling is to know how to handle buyer resistance.

Make connections. Initiate a connection something like this: I'm sure you're like me and are concerned for your family's health. I'm guessing you shop at the farmers' market because you want fresh, unadulterated, healthy food for your family. And you care about where your food comes from. Doesn't it make more sense to pay a little more for healthy food than pay later in doctor bills?

Nicki and Susan suggest using phrases such as these:

I appreciate your concerns. Or, I know what you

mean. Or, I see your point.

Repeat the objection and then use bridge words:

However, did you know.... or ...have you considered....

Warning Do not use the word "but," which has a negative and defensive connotation.

After crossing the bridge, tell them something good about your nutritious product.

Be prepared. Do your homework and write down any possible objection and a positive response to everything you can imagine. Price will be the most common objection.

Remain upbeat. Don't get defensive. And don't apologize for the price. Try this:

I know you can buy ground beef for 99 cents a pound at The Little Butcher Shop, however let me tell you why this is better....

Buying from us you know the farm and family this came from. You don't have to wonder if the animals were ever fed grain or spent time in a feedlot.

You want the healthiest food for your family—guaranteed to have no growth hormones, antibiotics, crowded conditions, or animal parts in their feeds. Our animals graze only fresh green grass every day of their happy lives.

"We don't take any government subsidies to produce our products."

Write down any objection you hear (hopefully by eavesdropping at the next booth rather than your own) and do some research if necessary to easily respond at the next confrontation.

If you're hit with a question that completely throws you, be honest: You know, I can't answer that, however if you'll be here next time I'll have an answer for you.

Scarcity. When he was selling at the Ferry Terminal Farmers' Market in San Francisco several years ago, Mark Keller discovered that scarcity would overcome price objections most of the time. His $26/lb organic beef tenderloin normally knocked people slightly off balance. Before they could regain enough composure to walk away, he would quickly add, "There's a limit of four per customer." Sold.

Others have found that having only one steak left in the cooler is easier to sell than a cooler so full the lid can barely close. Use smaller coolers and keep the extras in reserve. Smaller coolers are easier to pick up.

Make comparisons. Rather than selling by the pound, Morrilton, Arkansas, grazier, Ed Martsolf, who sells grassfed lamb at the Little Rock Farmers' Market, prices by the ounce.

Should you lower your prices?

Price is the basic way people initially determine value. Getting the price/value ratio right is very difficult and once established you must be very careful not to tamper with it too much. If you lower the price to move excess inventory, you will often find it very difficult to raise it back to its previous level.

"Prices paid today also influence consumers' expectations about future prices, and that in turn means that today's deep discount may end up eroding your future profit margins," noted Golden Gate University marketing professor, Michal Ann Strahilevitz.

It is actually far better to give excess inventory away as free samples than to sell it at a discount and ruin your price/value relationship. Free samples are absolutely the best way to build new customers for a food product with as much quality variability as grassfed meats. While farmers' markets are not a particularly efficient way to market meats, they are an excellent place to create new customers by using free samples.

Forty cents an ounce is not nearly as intimidating as $6.40 a pound, he said.

Since most people can't do the math in their head, he holds up a small Snickers candy bar and adds, It's roughly the same price per ounce as this candy bar. However, this grassfed lamb is high in omega-3 and CLA and is a lot better for you than the candy bar.

As Joel Salatin pointed out in a Polyface newsletter, "Is anyone angry that a BMW costs more than a Camry? Why do we assume we should have BMW food at Camry prices?"

Keep portions small. With over 40 years of marketing lamb, Martsolf has learned that products that cost less than $10.00 sell most easily. People come to a farmers' market looking for something for that night's meal. They don't come looking to buy a freezer-full of meat.

He wants everything he sells to fit into one hand so that he can hold the candy bar in the other for value comparison. As a result, all of his cuts are sold boneless, which lowers the yield, however, he doesn't worry about that.

He keeps his meat cuts simple because most are going to be grilled rather than roasted.

He encourages his regular customers to buy unfamiliar cuts by giving them samples.

All-inclusive. Let your customers know that unlike grocery stores, your prices are pure and simple. They don't have added fuel charges for transporting goods coast to coast. They're not covering employee wages. They're not paying for processed foods with ingredients you can't pronounce on the labels.

Finally. Someone who is truly objectionable and unpleasant isn't someone you want for a lifelong customer. If they don't like your prices or complain about blue skies on sunny days, and if they look like the kind of person who would kick a sleeping dog, keep your sanity and write them off. You'll sleep better for it.

Offering purchases for penny pinchers

To curb food costs these days, many people are taking the do-it-yourself route. They're becoming backyard farmers by growing vegetables at home. That's fine, but how many are capable of raising a cow behind the garage?

I wish I could take credit for that observation, but it came from Rosanna Bauman, a Kansas Rural Center Farmer Educator, who posts blogs for Kansas Farmers' Markets. It's a comforting feeling to know that as producers of quality grassfed products you can't be replaced by a do-it-yourself kit. Food is the last thing people will stop buying.

However, since people are more cautious about their purchases, your marketing has to be adjusted in order to appeal to penny pinchers. For the health of my family I refuse to substitute greasy grain-fed products just to have a cheaper meal. Won't your customers agree?

The price problem. Should you lower your prices in unsettling economic times? If you feel your prices are fair and reflect your goal of producing a sustainable income for your family, the answer as noted earlier is still "no." Discounts and sale items train customers to wait for the next sale before buying. Often "sale" signs indicate a business is in trouble.

Instead, promote your lowest-priced item.

If your prices were already competitive with supermarket brands, let your customers know this. Is there a Whole

Foods market in your marketing area? How about promoting your *Wholesome Food* at less than Whole Foods prices?

Point out that preparing a meal with even your highest priced item will be less than the cost of a restaurant meal.

Offer alternatives. Now is the time to introduce your customer to less well known cuts. Educate them on the alternatives to higher-priced ribeyes and tenderloin. For example, tenderloin tails are great for shish-ke-babs and just as tender and tasty at less than half the price per pound. Try bundling your cuts into Value Packs, including some of the cheaper cuts and hard to move items with somewhat pricier items.

Stress value. Promoting the value rather than the price helps customers justify their expenditures. Keep potential customers informed about the benefits of grassfed products over grain-fed.

Where's the meat? You can dramatically demonstrate the quality of grassfed beef by doing a comparison test with ground beef at your booth, or do it at home and photograph the results. Photograph pictures of the two uncooked items side by side or even better set each on a scale. Take a picture of each item being fried in a skillet. Drain the contents to show the difference. You might want to use a little rain gauge with inches marked on the side to collect the grease. The result—grassfed is meatier, grain-fed greasier. Plus, the grassfed product has more CLA and omega-3 than grain-fed. (Always stress those benefits.)

Farm-raised chickens typically have more meat per pound than the industrial variety. Bake one of your chickens and one of the other guy's, then weigh the cooked carcases. Be sure to photograph and list the differences.

Keep it simple. Make it easy for your customers to prepare a complete meal using simple recipes that you supply with every purchase. This is not the time to share elaborate recipes that require exotic—and expensive—ingredients. Keep the recipes simple—five items or less, suggesting spices and ingredients commonly kept on hand.

Promote the flavor of grassfed meats. We like steaks with only a sprinkling of salt for seasoning, unlike grain-fed meat that requires marinades, ketchup or steak sauce for flavor.

Meat stretcher meals. Occasionally even I blanch when I realize I'm cooking one meal with $30 or $40 worth of

Plumped chickens cost consumers billions each year

WASHINGTON: The Truthful Labeling Coalition estimates that American consumers annually spend an estimated $2 billion for added salt water in commercial grade chickens, *The Wall Street Journal* reported in 2009.

Currently, roughly 1/3 of fresh chicken sold in the USA is "plumped" with water, salt and sometimes a seaweed extract that helps the meat retain the water. And, the USDA allows such chickens to be marketed as "all natural" or "100% natural." This is hardly the case.

Untreated chicken has about 45 to 60 mgs of sodium per four-ounce serving whereas "plumped" chicken has between 200 to 400 mgs of sodium per serving. This is almost as much as a serving of fast food French fries.

Chicken companies say their private surveys show that consumers prefer the salty taste of the plumped chicken and said the practice was forced upon them by the large discount chain stores that wanted a uniform weight and price for each chicken.

A telephone survey of 1000 West Coast consumers found that 63% were unaware of the practice and 82% believe that salt-water injected chicken shouldn't carry the all-natural label.

meat. However, when I figure out how many people and meals that single entree provides, I average around $5 a person. That's more nutritious and cheaper than a fast food combo pack.

Create and distribute recipes centered around your grassfed meats that can be stretched for ample leftovers, which again will bring down the initial cost of an entree. For example, one of our favorite winter meals is Casoulet, a kind of French stew. It's chunks of grassfed beef, with sausage or ham, cooked in chicken broth with tomatoes, then several cans of beans—Great Northern, Black, Red—are added. I always use a lot of onions, celery and garlic. Sometimes when I'm reheating for the second or third time, I'll add another can of beans to stretch it further.

First timers. Offer "starter" packages in two or three price ranges to give the customer an opportunity to try your product. This could include educational materials such as the *Stockman Grass Farmer's* "Grassfed: The Informed Choice" brochure or a Q&A flyer, and recipes appropriate to the items in each package.

Play on their senses. Who can resist the aroma of grilling meat? Cook and offer samples. Cube the meat and stick a toothpick in each piece. Display it on an attractive platter. One taste says it all.

Satisfaction guaranteed. Customers reluctant to buy lamb (will it taste the same as beef?) or those who think all chicken tastes the same, may become your best customers if you offer to refund their money if they're not satisfied.

"Don't tell them you'll replace the meat," Rosanna Bauman warns. "If they didn't like it the first time, they probably don't want to try the same thing again." However, a backup warranty can tip the scales toward a sale. "This program encourages more feedback from your customers and gives you another opening for education, especially on cooking methods."

Marketing: make a plan

Creating a marketing plan

A marketing plan will only work if you stick with it. To that, write it down and post it where you can easily and frequently refer to it. Perhaps you'll write your marketing plan in the form of a check list so that you'll have things to check off as they're accomplished.

Niche identity. First figure out what niche you fill. Survey the overall market of competitors and look for a hole that no one is addressing. Perhaps there's a huge ethnic community in your marketing area and everyone is selling to them as if their products are generic items. Be the first to directly solicit those prospects. Research in order to understand their cultural background. Translate your product offerings and handouts into their language. Acknowledge their holidays or feast days with special attention. Ask for their help in assembling recipes. Learn their language or find a translator.

Goals. What do you want your marketing to accomplish? Increase name recognition? More customers (do you have the capability to ramp up production if you have a sudden increase in customers)? A way to move excess hamburger or other cuts with little demand? Become a grassfed leader?

Bull's eye. Once you've determined your goals, what do you need to do to accomplish them? Do you want to blanket your community with your brand identity? What kind of public relations can you do to be more visible? How can you change the difficult-to-sell cuts to make them more appealing or value added? Where should you expand your selling to reach your targeted market? What can you do to make your customers come to you (build that on-farm store)? Or should you take your products to your customers (establish a metro buying club)?

The plan. Write down all the different steps you can take to reach your goals. Prioritize the list by things you can do right now, things that will take more time, things you consider most important, and things you want to do...some day.

Budget. Determine what you can do for little or no money. Then attack the items on the plan. With more costly marketing ideas make a projection of its effectiveness. Include a means of monitoring results. Make sure your best ideas are working for you, especially when there's a cost involved.

Assignments. Decide who will do what, and always give them a time limit. Make it their responsibility to track results when they report back to you.

The test. Go into action. Get the plan working for you.

The results. Track the results of each aspect of your marketing plan. Review the outcome quarterly, then restructure the plan accordingly.

Some thoughts on marketing goals:

- Think of ways you can better serve your customers.
- Can you add more products to those you already sell?
- Is there an inexpensive or free service you can add that distinguishes you from the competition?
- Are there systems or equipment that could help you achieve your goals—a computer upgrade, copy machine, accounting software?

Once you've written out your marketing plan, put it into action, and gotten the results of at least one cycle of its use, it's time to write another marketing plan. Multiple marketing plans can work to increase profits just as multiple products can increase revenue.

■ Each plan can be directed toward different targeted customers.

■ Having multiple plans allows you to recycle them to keep things fresh and different without having to dream up a new approach every time.

■ Whenever you add something new to your marketing plan consider dropping something that's ineffective. Always track results so that the weakest link can be eliminated.

■ Don't ever toss old marketing plans. After being out of use for a period of time you may re-discover them as best sellers in different times.

21 questions about your competition

In preparing your marketing plan, do your homework by checking out the competition.

1. Who is selling the same thing you are? Do you have one competitor, or many?

2. What is their competitive advantage? Try to identify the superior aspects of their marketing. Perhaps they're selling more than you because they have multiple venues—farmers' markets, on-farm store, buying clubs. If growth is a goal, figure out how you can expand your sales locations.

3. What can you learn from their success? You will have to figure this out by observation as very few competitors will share tips or practices with the competition. It might work once to be a "spy" by pretending to be their customer, but don't jeopardize your relationship just to steal their secrets.

4. What is the most obvious thing they're doing wrong? This will most likely be the first thing you notice, because you'll be comparing what they do to your own practices. Ask yourself if you noticed that flaw because it's one of your own failings. Be honest.

5. What do they do best? This too should be obvious, however, their excellence can be intangible. Perhaps they provide super service or customer relations that have customers flocking to them. Passion alone could be the difference between something that sells well and something that's just on offer to

someone who might look your way.

6. Why do customers seek them out? Showmanship can be entertaining and garner interest, but is it genuine? A good show won't bring out people's wallets unless there's a quality product behind the hype.

7. Why do customers shop with them? Perhaps their prices are lower or their signs are pretty. Look beyond the cosmetic appearances to discern the heart of this answer, then figure out a way to improve your own marketing in a different way.

8. How do your prices compare? For some customers, shopping is a game. They enjoy the hunt for bargains and like to carry fistfuls of coupons. Others like to sample everything before making the decision to buy. Search for reasons your competitor prices things differently than you, and don't lower your prices thinking that will work to your advantage. If you priced your products correctly initially, you shouldn't compromise yourself by lowering prices just to make a sale. It will hurt your bottom line in the long run.

9. How many products do they have? Let's say you are both selling grassfed beef. You're selling quarters, but they're selling by individual cuts. Or perhaps they're adding value by creating pre-packaged bundles. Figure out ways to repackage and multiply your product offerings.

10. If you're both selling ground meat, how do they manage to move all of it when you're left with mounds unsold? Are they adding value by selling cooked burgers? Are they packaging ground meat in with other pre-selected cuts? Are they adding a strip of bacon and calling it gourmet ground? What's holding you back from making your products more diverse?

11. Do they have a better product than you? This is a reality check. Be sure you are producing a quality product. If your product needs improvement, you should retreat from sales until the quality improves. Customers who have a mediocre first experience with grassfed products will not only boycott the

vendor who sold it to them, but will give up on grassfed altogether. You owe it to yourself and the grassfed industry to produce the very best.

12. Are they organized? At a farmers' market are they already set up and selling before you can get everything off the truck? Identify what makes them so efficient—do they have a system they follow every time, or just a big, burly helper who lifts weights? Take time to write down everything that needs to be done before the first sale is made and then prioritize steps. Write this down and post it where you can follow this every time to be better organized.

13. How often does your competitor's name come up? Promotion is a never ending project. To create word of mouth—once you have a quality product—you have to give people something to talk about by being written up in the local press, sharing grassfed tips on talk radio or TV, or sponsorship.

14. If you weren't selling the same thing, would you buy from your competitor? Regardless of whether you answer yes or no, ask yourself why.

15. Are they selling things other than grassfed products?

Things your mother taught you

■ To appreciate a job well done: "If you're going to kill each other, do it outside—I just finished cleaning!"

■ Religion: "You better pray that will come out of the carpet!"

■ Logic: "Because I said so, that's why."

■ Stamina: "You'll sit there until all that spinach is finished."

■ How to solve physics problems: "If I yelled because I saw a meteor coming toward you, would you listen then?"

Internet posting

Take inventory of the variety of things they have to sell. Ask yourself if these items are related to the core products. Brainstorm like-minded goods that you could add without duplicating the things they sell.

16. Who is selling? An outstanding "people-person" personality will be hard to duplicate. The easiest sales person to trump is someone who is "sent" to cover the cash register. A person without passion or an investment in the product, who is just doing a job can be detrimental to sales. They don't have informed answers to questions. This undermines the trust previously built between seller and consumer. Such a person may be able to complete a sale, but will hardly encourage new customers to return.

17. What do their sales people look like? Are they dressed neatly and attractively? Are they wearing some kind of uniform, even if it's a polo shirt with their logo? Does what they're wearing stand out?

19. Are they capturing customers' names? Someone collecting business cards or names, perhaps with a giveaway as enticement, will be creating a database for future, expanded sales.

20. Do they do follow up after a sale? It may be difficult to determine if they stay in touch with their customers. However, if they're collecting names, you can be sure they're putting those names to use.

21. Are they pursuing all of their potential customers? Perhaps there's a demographic or ethnic market they're overlooking. Seek ways to fill those gaps.

Once you've identified areas that need improvement, talk with a similar vendor who is not in your selling area. Make friends with like-minded marketers in another state or region. They'll be more likely to share information when you aren't going head-to-head with them on sales. Learn from your mistakes. Make a commitment to improve one thing at a time.

Touting your message

When a man sees a pound of ground meat he thinks *hamburger*.

Women think *chili, meatloaf, sloppy joes, tacos, meatballs*.

Women are not necessarily visionaries, we're just more sensitive to other people's needs. In a sales context, we're simply helping people discover *if* they need what we have to sell. You want to give them plenty of choices to convince them that *your* product is what they want to buy.

Begin by defining your sales message before the sale.

Tell your story. Californian, Barbara Bowman, who sells specialty meats in the San Francisco Bay area, says you have less than a minute to use your story as a hook. It must fit you like "an old, worn shoe." In her experience, you will get tired of your message just when the consumer is starting to hear it. "You really have to have passion to persevere," she said.

Allan and I love telling our friends that the vegetables we buy come from a burned-out Berkeley, California, professor who moved to Mississippi in order to grow organic produce. What kind of story is there behind a Wal-Mart tomato?

Many of you have exchanged fast-paced urban careers for a pleasant life in the country. Others have come to grass finishing for health reasons. Some of you are third, fourth, or more generations of farmers. Sharing a little about your background will help your customers remember you and make their purchases more personal, meaningful, and real.

Let your message reflect who you are and what you represent—a healthy, happy lifestyle in sync with nature.

Use an anecdote. At the *Stockman Grass Farmer* we like to ask where subscribers heard about us. My favorite story came from a farmer who was on a ladder painting his house when he heard Allan speak on Howard Garrett's Sunday morning Dallas radio show. When Allan gave the magazine's toll free number the man painted it on the side of his house.

Everyone will bring different ingredients and styles to their sales message. While you may want to script your message initially in order to feel more comfortable, be spontaneous and flexible to fit the situation and keep your message fresh. Have two or three messages to use for different people and occasions. You don't want to sound like a robot delivering the same script to everyone.

Emphasize benefits. Organize benefits in terms that would sell *you.* If your products are seasonal, make that part of your message. Or how about "Not available in stores!" Health-conscious consumers will want to know about the higher CLA and vitamin content found in grassfed products.

Do you have a money back or exchange guarantee? Why not offer a Ready-for-the-Crockpot Roast and include a zippered bag of spices and a recipe? Or Packaged for Two.

Write down all the key points about your product or farm so that you'll be prepared for any question the customer might ask.

Always be looking for new ideas to give your customers reasons to buy.

Include visual aids. Let your farmers' market booth or on-farm store help you sell. Create posters or free handouts listing health benefits. Display them as FAQs (frequently asked questions). Offer brochures about your farm, your family and your products.

When he was selling at farmers' markets Joel Salatin's

Polyface Farm booth at Dupont Circle in Washington, D.C., was cleverly decorated with green fringe that at first glance looked like grass, but a closer look revealed the cutout shapes formed forks and knives.

Collect articles on grassfed products in an album for customers to browse. Tout your successes and good press. Laminate newspaper clippings and hang them from a clothesline. Physical activity gets customers involved. Have a flip album of farm photos, or put photos on a revolving stand that spins. Leave a scrapbook with farm press and articles on health open to a significant page.

Remember, these are support materials. Handouts don't sell. You do.

Testimonials. Collect customer testimonials and include them in your brochures or signage. Ask your best custom-

Rolling marketing

Randy Sipe of Sipe Angus Farms in North Carolina took a white panel van with a side-awning flip out and had it plastic wrapped in signage. The background is a lush pasture of Angus cows. In addition to his phone number and email address, he listed these benefits:

Ingredients: Grass, Water, Sunlight.

Proven to Contain Omega-3s.

It's Heart Helpful.

No Hormones or Antibiotics. Humane Farm Animal Care.

Environmentally Sustainable.

He takes this rolling restaurant to outdoor events where he and his family cook and sell grassfed burgers and country fried steak. His sales message is working anytime the vehicle is on the road.

ers to write one for you. Paper the wall of the farm store with them. Place them under glass on the counter with the cash register or decoupage them onto picnic tables where your customers dine at your farm store. Print them on your invoices and order forms. Put them anywhere someone is likely to wait. Wintzell's Oyster House Restaurant covers the wall with one line sayings. It keeps diners busy while they're waiting for the food to arrive...and gives those couples who have nothing to say to one another something to do.

People who are undecided are uneducated about your product and need more information. When you refer to a customer testimonial it helps defuse a prospect's skepticism. Invite them to visit the farm (and to bring a friend) to see for themselves everything you tell them about.

Seeing is believing. A Northwestern University study found that graphics or pictures added to a written or spoken message increase recall 300%.

Use photographs, paintings of your farm, or graphics. Your children's drawings of various aspects of farm life or animals can charm customers to come closer and ask more questions. How about including a banner sign that reads, "Thank you for supporting the renewal of the land and the lives of the people who live there."

Be sure everyone is on the same page. On one of our vists to Whole Foods we couldn't find brochures or branded grassfed beef products in the meat case, so we asked the person behind the counter where the meat had come from.

"Canada," he quickly replied.

"Canada?" We parroted. "That doesn't sound right. Isn't there an embargo on their beef?"

He ran off to check with someone else. When he returned he said, "It's from New Zealand."

So much for a policy of local sourcing, but that's another story.

The point is, if someone is helping you, be sure they know your story or can at least direct the customer to the proper brochure or sign to answer questions if you're occupied elsewhere. Your on-farm store or farmers' market booth should be a creative, attractive environment with upbeat, smiling people eager to help your customers. If your children are helping, be sure they enjoy being there and being part of the action. Give them a special job—setting out and keeping brochures straight and neat, or wrapping/bagging purchases.

Freshen your message. Update your message periodically so that it won't become bland to you or your customers and no one will take it for granted. Fresh ideas give existing customers reasons to buy again. You should constantly give your customers reasons to buy more and to buy more often.

Constantly remind consumers who you are and what you do, and always add something new. For example:

We're 100% grassfed, and healthy for your family.

We're 100% grassfed and on the Internet.

We're 100% grassfed and accept credit cards.

We're 100% grassfed and can now ship to your out-of-town friends.

We're 100% grassfed and we want to be *your* farmer.

Make a big deal out of every "new" thing.

Put a new spin on old sales messages. Brent Walter, marketing and sales director for Fox Fire Farms in Ignacio, Colorado, issued a "new" press release titled, "Colorado Farm Switches to Argentinean Farming Techniques to Offer More Wholesome Products." No one has questioned the fact that they're still using the same (Argentinean/North American) Management-intensive Grazing methods as always, but have added an Argentinean style forage-chain of grasses to extend grazing year-round.

Never assume you've done all you can and have reached the top of your market. There are always things to tell to make people relate to you and want to buy your products.

How to sell: putting the customer first

Who are your customers?

Have you given much thought about who your customers are?

Not just their names, but who they really are, where they live, what their interests are, and how many mouths they have to feed.

Although I'm not advocating putting labels on everyone you meet or forcing your customers to fit inside a neatly packaged box, categorizing them will give you marketing clues so that you can better serve their needs.

Marlene and David Groves of Buffalo Groves raise buffalo some 75 miles southeast of Denver. They wanted to sell their products outside of supermarkets and traditional restaurants where the "big guys" sold their meat. In searching for the best farmers' markets in their area they ignored the usual profiling of young, old, single, married, male, female, working executives and retirees. "Let's face it," Marlene noted, "they all eat."

Instead she looked for customers who were Health Minded, Wanted Local, Wanted Natural, Had Money to Spend, and Had Time to Cook. The Groves realized that people who fit that profile would be shopping for quality, not quantity. Those customers bought locally because they wanted to know more about the products (and producers) they were purchasing. They also enjoyed relationship—buying from someone they knew.

Doug and Krista Dittman have a small organic raw milk dairy near the university town of Lincoln, Nebraska. The sale of raw whole milk draws customers to their farm, which encouraged them to offer raw milk cheese as well.

They identified their customer base this way:
1. Youngish Baby Boomer grandmothers concerned about their grandchildren's health. 2. Home schooling Moms. 3. "Picky but not gourmet" consumers who want locally produced organic food. 4. People with health problems.

Home schoolers are among the customers of Joel Salatin's Metro Buying Club patrons. Other interests of these Metro ladies are that they are fundamentalist Christians, Jews

Cream of the Crop

In *The Purple Cow*, Seth Godin says direct marketers have only four kinds of buyers—Prospects, Customers, Best Customers, and Former Customers.

Your Best Customers should be thought of as an entirely different category from occasional customers. Kansas dairy grazier, Kenneth King, has created a Cream of the Crop Club for customers who have purchased over $1000 within a year.

"These customers will receive advance notice of new products and services, and when products are running low," he explained. "In addition, Club members will receive other special rewards throughout the year."

Customers are notified by email when they qualify. He emails: "A BIG thank you to our loyal customers " and adds, "Would you like to be in the Cream of the Crop next year? You know what to do."

following *The Maker's Diet*, Weston A. Price members, and members of Families for Natural Living. Many are soccer moms.

A large number of Joel's buyers are conservative Catholics who have pulled out of the Catholic school system and started their own school. "The common factor among all of our customers is that they have decided to opt out of the system," Joel said. "Most are ex-Whole Foods customers who have become disillusioned with the trend toward industrial organic."

What can you learn from these customers' interests?

Healthy food is their number one concern. They care about products from animals that have led a healthy lifestyle, as Jo Robinson of eatwild.com has said, "You are what your animals eat." Once they've been educated about the greater healthy attributes of grassfed products they won't settle for store-bought grainfed anything. They want more than just the illusion of healthy products bearing meaningless labels such as "fresh," "natural," "real," "organic," and even "grazed on pasture." They want to know they're getting 100% grassfed milk and meat. No compromises.

Educated. The fact that they home school doesn't mean their kids are smarter, but it does signify that they know enough to seek more than the norm offered by conventional schools, supermarkets, and society. They read. They ask questions. And the answers they've received lead them to seek something "beyond organic."

These will be your best customers. Not only do they understand and appreciate hormone and antibiotic free, unstressed animals, healthy green pastures for the entire life of the animals, chickens that graze outside in green pastures with nature's bounty of bugs, but they will educate their friends on the finer points of healthy eating.

Busy lives. Just the fact that Metro Buying Clubs exist suggests that the people in them have busy lives. They love saving time by having their purchases delivered to them, and are willing to pay more for that privilege. Since their time is precious, it's important to respect that by always being on time for deliveries or keeping regular on-farm store hours as posted to fit your own schedule. They're virtually pre-sold, so just fill up their shopping bags.

Have money. While many of your customers may not be affluent, when they care about healthy food purchases cost is not a leading motivator. You'll know who these people are—as the ones who buy a multi-month supply of something rather than "just one to sample." The samplers may eventually turn into regular big buyers, but the point is they're not afraid to splurge on food purchases for their families, so don't feel cowed into pricing yourself out of the market to save a few pennies for your customers.

Have time to cook. These customers enjoy the process of preparing meals for their families and friends. They will splurge on high priced cuts for entertainment. They like to experiment in the kitchen and will eagerly accept new twists (European style butcher cuts, bacon-wrapped-filets) on the same product. These are not your average hamburger helper cooks.

Relationships. Who you are and what your products represent matter to them. These are not fly-by-night buyers. Once they discover the healthy, wholesome products you sell and trust your exceptional products and services, they'll be customers for a long time to come.

Dr. Steven Atchley's consumer group test identified three types of customers for grassfed beef: Baby boomers who were fighting aging and wanted to live forever; young mothers concerned about the effect of nutrition on their children's health and academic achievement; and Generation X and Y consumers in the 18- to 30-year-old age group who have a social consciousness about what they buy. "Educated consumers know the difference and are willing to pay a premium price for it," he said. "All we have to do is to help educate them."

20 questions about your customers

How well do you know your customers? Take this quiz to learn more about them. Make your best guess from your observations as you interact with them.

1. Where do your customers live? Get their address for your database and check it out.

2. Are they from upscale subdivisions or live in kid-friendly neighborhoods? People living in retirement communities will have different priorities from families with school-age children.

3. What's their level of income? Where a person lives often reflects their level of discretionary income, however, don't be fooled. Often that upscale subdivision has plenty of houses for sale and cash-strapped residents who care about outward appearances more than real values. We discovered this years ago when we were house shopping and found many who lived in pricey homes couldn't afford to furnish them. They were constantly moving because they couldn't keep up with the notes.

4. What kind of car do they drive? This is another value indicator, but again looks can be deceiving. Factor in this observation with others on this list.

5. Do they budget their purchases by the amount of cash they carry with them? This could be a good sign that they care enough about healthy food choices to spend as much as they are

able. Some customers will pay cash because they are aware it will save you the percentage lost from credit card purchases.

6. Do they always pay with a credit card? This may be for convenience sake, or it could be a sign they are near their credit limit.

7. Do they treat you like the family counselor and talk about their problems with you? Sharing incidents in their lives is a bonding process. It shows they trust you and consider you somewhat of an expert about your products and healthy foods.

8. How and where do they learn about food products? People who read a lot, particularly about health issues, show they are open to learning what you have to say about your grassfed products.

9. Do they pay attention to your signage or pick up flyers or brochures you've prepared for give-aways? This is another sign that they are open to buying from you.

10. Do they ask a lot of questions? Often this is a sign of an educated consumer who doesn't take things at face value. They always want to know more. It could also indicate that they aren't 100% sold on what you have to offer. Ask yourself if there is something about your marketing that could be confusing to the customer and then fix it.

11. Where do they shop, other than with you? Pay attention to what they're wearing - designer chic with everything coordinated and accessorized, or T-shirt casual. Have you noticed them wearing the same outfit more than once?

12. Are their farmers' market purchases minimal or do they buy a variety of goods? This is not a sign of someone preparing for a party. Someone who purchases a lot of things at

Who cares?

"If you aren't taking care of your customers, your competition will."
Albert E. Schindlee

a farmers' market probably minimally shops in supermarkets. That's good for you.

13. Do they shop with you regularly? Frequent purchases indicate satisfaction with the product and a commitment to you as their personal farmer.

14. How concerned are they about the environment? Customers will choose your products over someone else's if yours reflect good environmental practices such as clean water and lush green pastures for your animals.

15. Do they carry a reusable shopping bag or basket? Their concern for the environment means they will seek out like-minded people for their business transactions.

16. What are their interests, concerns? Soccer moms are pressed for time, so are moms with many children.

17. Does "local" and "seasonal" matter to them? They most likely shop with you because they want the freshest products for their families. They don't want a baseball-hard tomato from Chile in the middle of winter. They stock up on products at their peak and practice freezing and canning to tide them over out-of-season.

18. Do they ask where your products came from? Your signage should answer this. They want to know they are getting genuine products from the farm family who produced it instead of getting a second-hand sale from an unknown source. This applies to their fruits and vegetables as well as their meat and milk.

19. Do they tend to avoid shopping with vendors who purchase goods wholesale just to resell at the farmers' market? This is best judged by how well the other guy is selling. Some people don't care enough to ask. Those who do are the kind of savvy shopper you want to engage.

20. Do you have a history together? You will if you stay true to your product and give your customer all the time and answers they need before making that first purchase. You will be rewarded when they come back again.

Once you've answered these questions and have a better

understanding of your customers, find ways to meet their needs with your products and services. There are exceptions to every situation, no hard, unwavering answers. However, you'll be a jump ahead of your competition when you know more about the people who come to buy from you. This separates the amateur marketers from the professional.

Listening

Three retirees, each with a hearing loss, were playing golf. One remarked to the other, "Windy, isn't it?"

"No," the second replied, "it's Thursday."

The third chimed in, "So am I. Let's get a drink."

Listening to your customer is key to making a sale.

Since women are more likely than men to tune in to another person's emotions and body language, it follows that we're better listeners. Aren't we the ones who always have an ear cocked for a crying baby (or, in our household, the sound of our dog, Rascal, ready to bury his head in the garbage can)?

Those who have mastered listening skills can help direct prospective buyers into discovering what they truly need, even when they may not realize it themselves. Women communicate by telling their life stories. In other words, good listeners will pick up on how they plan to use your products, and whether they are for herself, her family or her friends.

It's not about you. It's all about them. Focus on your prospect. Be an active listener, engaged in what they have to say. They don't care what you have to say until they've had their say. Maintain eye contact and smile. Nod your agreement. Let them take center stage.

Ask good questions. While it's good to be prepared for what you *might* say in various situations, never use a canned speech for everyone.

Start by asking good, open-ended questions—the kind that encourage discussion. Open-ended questions allow you to customize the exchange so that you can respond to each prospect's needs, wants and worries.

Avoid questions that only have "yes" or "no" for an answer. They're dead ends. When given the choice, most people will say "no."

I learned this the hard way early in my writing career. Mohammed Ali was in Natchez, Mississippi, filming "Freedom Road," and I was allowed to interview him during a lunch break. (Coincidentally, Allan was there too, to get his own interview, but we didn't meet until years later.) Like the professional I pretended to be I flipped out my notebook, poised my pen and asked this boxing champion my best, and only question, "How do you like acting?"

"I like it just fine," he replied and went back to eating a plate full of macaroni and cheese.

Stuck without a follow up I turned in a story that heavily featured what he ate at lunch.

Here are some examples of good, open-ended questions:

"What are you looking for when you're planning a meal?"

Improving listening skills

■ Make eye contact with the speaker.

■ Focus on what they're saying, not how they speak.

■ Don't let your mind wander.

■ Treat listening as a challenging mental task. (Pretend you'll be tested on what they say.)

■ Active listening means asking yourself mental questions.

Adapted from *Student Handbook: Listening Skills,* University of Minnesota Duluth.

"What kind of meal are you planning—a special event, or a meal to stretch your food dollars?"

Offer pros and cons of various products so that she can feel she has considered all the options in her decision making.

When asking questions, pretend you're on a first date with the customer to draw them out.

Silence. Skilled interviewers know that silence (listening) can draw the other person out. Silence makes the other person uneasy, and they will often do all the talking just to fill the void. Well, maybe everyone but Ali.

"The single most interesting thing you can do...is to ask a good question and then let the answer hang there for two or three or four seconds as though you're expecting more," said newsman Mike Wallace. Although he was referring to television news interviews, it works in selling situations as well.

While you're being silent, you can read between the lines. What can you learn about the other person by the way they act, speak, dress? Listen with your senses and your eyes by maintaining contact. Mirror their actions—frown with them, smile with them, laugh with them.

Listening (with a smile) shows interest on your part. It's respectful and imparts a sense of being valued. Silence can send a healing message. It can convey understanding. At the same time it encourages the other person to open up and share their wants, needs, experiences and expectations, all of which should give you clues to guide the conversation toward a sale. Sometimes what the prospect is really saying is never verbalized.

Above all, avoid distractions. Clench your hand or curl your toes if necessary, but don't show your own nervousness by playing with a pencil or checking your watch. Interrupting can make the prospect lose their trend of thought, and it's impolite.

Attentive listening. By giving the prospect your rapt attention, you show involvement. For this reason alone it's important to have help so that the helper can continue with the necessary busy work or fill someone else's order while you focus on the prospect.

When Joel Salatin makes deliveries to his Metro Buying Club members he never goes alone. This allows Joel to chat with his customers while the other person unloads the orders, collects the cash, and watches the clock to keep them on schedule.

When you are actively listening you have to resist the temptation to interrupt the speaker.

Lean in, which is body language saying, “I’m listening.”

Repeating or rephrasing something the prospect said is another way to show you’re involved.

Clarity. When it’s your turn to speak, be ready to clarify what the prospect said, by saying something like, “Am I correct in assuming...?” or “Do I understand you to mean...?” or “What I believe you’re saying....”

Then follow up with suggestions to fit the situation.

Being time conscious. Customers always appreciate the courtesy of respecting their time. Be prepared to say, “I know your time is precious, so let me just say this....and here’s a brochure that will tell you more in detail.” Then quickly promote the top three value benefits of your products.

Giving thanks. If you don’t make a sale after the first conversation all is not lost. Your prospect may just be too busy to make a decision now. Assuming she is still interested, say again what’s of benefit to her, and thank her. Being thanked when she hasn’t bought a thing and isn’t expecting it creates a surprise. This can cause her to rethink and purchase another time, and lead to positive word-of-mouth recommendations to the people in her network.

Finally, be yourself. The more comfortable you are, the more easily you’ll draw customers to you time and again.

How to get customers to sell themselves

Wouldn't you love it if you didn't have to do a thing to sell your grassfed products? Wouldn't it be wonderful to just stand back and watch your products walk away with satisfied customers?

Let's turn the tables and have your customers sell themselves.

Ron Willingham writes in *The Best Seller! The New Psychology of Selling and Persuading People* that prospects who sell themselves have fewer objections making it easier for you to close a sale. It starts with basic communication skills: make eye contact to focus on them; make them feel important by getting them talking about themselves; and ask insightful questions so that you can match your best products with their needs.

Open-ended questions. Establish rapport by getting to know them. Ask about their family. What kind of activities are their kids involved in? Are they newlyweds? How long have they lived in the community? If they're newcomers, what brought them here? Be sure to make notes after the sale so that you can pick up where you left off next time they come to buy.

Listen. Keep the focus on the prospect. Show them you are listening by rephrasing what they said or asking for clarification. "Listen" for their feelings: are they defensive, unsure of themselves, in a hurry? They will be more likely to open up if they sense you care about what they say. Agreeing with them—"You're so right," "That's so true," "I agree,"—lets them know you're listening, and keeps them talking.

Identify their needs. Your questions should help you learn about their needs. How do they plan to use your products—to feed a horde of visitors, to impress a date? Be careful to not come across as prying. Instead give the impression you are partners in their selections. Take a reporter's approach to ask the Five Ws—who, what, when, where, why. This helps them to make up their own minds when they're uncertain of what they want or need.

Be indirect. Don't ask them what they want to buy. Instead, figure out together what they need and then suggest the best buy solution for their specific situation. Often prospects don't know what they want or need. Your open-ended questions should lead them to draw their own conclusions of what will work for them. Pretend you're a veterinarian and your pet prospect can't tell you what they need. Look for revealing symptoms by observing their body language to put them at ease. The longer you keep them talking while you listen, the easier it will be for them to sell themselves. The more they talk, the easier it will be for them to trust you and your products.

Clarify. Repeat a question to clarify what they're saying. Periodically make sure you're both on the same track. As Willingham noted, "Asking questions such as these elevates you to the role of a sales counselor rather than a slick, impersonal operator. It sets you apart from salespeople who don't really care what the prospect buys as long as he or she buys something."

Questions repertoire. When you ask a question that draws a great response, remember it. Put it in your arsenal of questions to use in other situations.

Demonstrations. The use of demonstrations can often overcome price objections. However, don't start a demonstration until you've established your prospect's needs. A demonstration makes a positive statement to reinforce the benefits of your products and how that will meet your customer's needs. For example, someone who loves baking can't help but be impressed by cracking open one of your eggs next to the watery generic egg from a supermarket—yolks are orange, the raw egg is "heartier" (and healthier), which makes cakes richer and more substantial. Or fry a bit of your ground beef and the store variety to demonstrate the excess grease of the supermarket brand. Doing the demonstration together will have more impact on the prospect than those sales people who stand in grocery stores with product samples hoping someone will stop by.

Push the emotional buttons. Successful sales people have learned that tapping a prospect's emotions has better results than appealing to their logic. Sure, there are practical reasons that make sense for buying, but the emotional reasons of keeping families healthy can tip the sales scale. Help your prospects see the rosy red cheeks of their children, giving them healthy meals and snacks to satisfy their appetites without contributing to the nation's obesity problem. Tasting a sample can add to this emotional impact. Remember, your products are going to make them feel good, be healthy and look good.

Power positions. Where you stand can subtly affect buying decisions. If your prospect is taller or stands slightly higher up than you, that puts them in the power position. When they're in a dominant position you and what you say are not a perceived threat. Studies indicate that men selling to men

Listening tricks

When you make eye contact with a prospect, focus on just one eye rather than both. It may seem odd at first, but practiced for a few weeks it will become a habit.

Become a value generator by selling something to someone that's more valuable to them than money. Make people feel valuable by your vested interest in them.

should keep an arm's length apart. Women selling to women can close that gap a bit. Right-handed buyers want the sales person standing just to their left. If you're seated behind the sales counter, use a stool that has a high position so that you're almost standing, but not quite.

The price question. Often the question of price will come up sooner than you want it to. You can distract or defuse this situation by saying, "Let's figure out what best fits your needs before we explore the options." Then always emphasize the value of your products. Keep the questions coming: "How do you feel about this?" "Which package do you think would work best for you?" "In your opinion, what works best for you?" Always offer, "Once we decide what you need we can always take something out to reach the price you want to pay."

Negotiation. Whatever you do, don't lower your prices to make a sale. Don't get defensive or apologize for your prices. Don't use your competition as target practice. Take a deep breath and focus on the customer's feelings. Say things such as, "I understand how you *feel*." "Other customers had similar concerns, however most *found* this was negligible compared to the healthy benefits for their families." Willingham suggests asking, "Is there anything that's keeping you from buying now?" Focus on value. Be ready to re-state the benefits of your products and service over your competitor's.

Make friends. Focus on developing relationships with your customers. They'll become your best advertising by word of mouth. They will trust you and your products. And they will become lifelong friends. Everyone finds it easier to buy from a friend. To help you win someone over ask for their opinion, advice or help on a subject totally unrelated to what you're selling. It makes them feel important and valued. Look beyond the success of the sale and see the benefits a long-term relationship can bring.

Willingham advises, "You don't *talk* people into buying, you *listen* them into buying!"

Viva la difference!

Here are some basic characteristics of men and women. Savvy marketers will capitalize on these differences when it comes to selling.

Soft. Women have what Faith Popcorn, author of *EVEolution*, calls "peripheral vision." We are swayed by relational elements. Does the booth have pretty pictures of the farm family? Is there a vase of flowers that the daughter picked from the garden, or even better an unusual bloom that can be a conversation opener? Can prospective buyers munch on home-baked cookies while deciding on their meat purchase? Is there a tantalizing aroma of grilling meat to sample?

Hard. Men are accustomed to direct, hard-sell situations. (Could this be where the term "bull-headed" comes from?) They like things short and to the point. As to details, put those in an instruction sheet. They're not going to read them anyway. (How many of you women have ended up assembling your child's first bike?)

Ask. Women are not afraid to ask for more information, or directions. Knowing that, Allan always asks me to give first-time guests directions to our house. I'm generally the navigator when we travel. (Frankly, I have no clue how Allan finds his way home when I'm not traveling with him.) Sometimes when we're together in the car, I don't say anything when he should be making a turn because I want to see where he'll end up without my help.

Glinda, our partner, said her husband got on the wrong commuter plane once when she wasn't traveling with him. Let me add, he's a pilot.

Any kind of additional information you can hand out will add value to the sale.

Don't ask. Men would rather go into their "caves" rather than openly talk through a problem. They like to concentrate on one problem at a time. If they help in the booth, let the man do the grilling while the woman does the talking.

Women

- Women value love, communication, beauty and relationships.
- Women spend a lot of time supporting, helping and nurturing others.
- Women define their sense of self through feelings and relationships.
- Women care more about relationships than work or technology.
- Communication is job one for women. They feel it's more important to express their feelings than to strive for success or arbitrary goals.
- Women intuitively can anticipate the needs of others.
- A woman's sign of love is to offer help or advice without being asked.
- Because proving one's competence is not as important to a woman, offering help isn't offensive. Women don't regard the need for help as a sign of weakness.

Connect the dots. Women are relationship oriented. Within minutes of meeting, women will ask about children, schools, and spouses. They will make connections to mutual interests and friends. They'll swap recipes, tips, recommendations. "Women are three times as likely as men to recommend brands when they know friends are looking for a particular product or service," Popcorn wrote. They need a lot of information when making important buying decisions, and will appreciate it when someone has everything clearly spelled out to aid in that decision.

Many men perceive women's chit-chat as distracting and time consuming. What they don't realize is that talk is part of a woman's bonding process. When women feel connected,

Men

- Men value power, competency, efficiency and achievement.
- Men do things to prove and develop their power and skills.
- Men define their sense of self by getting results.
- Success and accomplishment make men feel fulfilled.
- Men are most interested in whatever can help their achieve their goals.
- Men like to do things by themselves. Being self-sufficient is a sign of their power and competency.
- Men don't give or graciously receive unsolicited advice or help.
- Being asked for help or advice gives men a sense of validation and competence.

Men are from Mars, Women are from Venus, John Gray, Ph.D.

they're more likely to trust what the other person has to recommend. And buy.

If your spouse is in the booth with you—and thanks guys, we appreciate your heavy-lifting abilities—let your spouse know that when you're talking with a customer, you're working on your friendly, informational sales technique. If there's a technical question you can't answer, defer to him.

Arrow-straight. Men are more direct. What's the point, they want to know? And let's get to it. That approach might work if you're the only one selling meat in a farmers' market. But if there's competition, you have to get creative.

Don't tell. Women will rarely tell a man what they want. As nurturers, women can anticipate the needs of others. It's second nature for us to pay attention to body language and emotional, unvoiced, messages. If a woman has a bad buying experience, she will rarely tell you. She just won't go back. And when she's really disappointed, she'll tell all her girlfriends, who won't go back either.

Tell. Men just want us to tell them what we want. What we want is to be allowed to trust our instincts when selling at the farmers' market.

Words at face value. When a woman says, "I'll think about it," that's precisely what she means. She needs a little extra time before making up her mind.

The message behind the words. When a man says, "I'll think about it," it means he's not interested.

Nodding women. When a woman mumbles, "mmm-hmmm," and nods her head, stop what you're doing. You've lost her. Ask a few questions that allow her to give you some feedback in order to get back on track. For women, head nods alone don't mean she's buying your message. It only means she hears what you're saying, but may not agree with you. Out of politeness, she's just not stopping you.

Nodding men. Men on the other hand, are agreeing with you when they nod their heads.

Minimalist. A savvy marketer once noted that men keep six things in the bathroom—a razor, shaving cream, toothbrush, hair brush, soap, and towel.

To the max. Women keep 371 things in the bathroom, most of which men can't identify.

Tuned out. Men have a tendency to focus on one thing at a time. They're single-minded and straightforward—good traits, especially in time-critical situations. Give him just one selling point to concentrate on at a time.

Tuned in. Women excel at multi-tasking. We can be planning the evening meal, totaling a customer's purchases, and sharing cooking instructions at the same time. So, step back guys and let us do the marketing.

Have you ever wondered what might have happened if there had been Three Wise Women instead of Three Wise Men?

They would have asked directions, arrived on time, helped deliver the baby, cleaned the stable, made a casserole, brought practical gifts, and there would be peace on earth.

Guidelines for men selling to women

A man is walking along a beach when he notices a bottle that has washed ashore. He picks up the bottle and out pops a genie.

"Thank you for rescuing me," the genie says. "I'll grant you one wish."

The man rubs the stubble of his beard and muses, "I've always wanted to go to Hawaii, but I'm afraid of flying. Could you build me a bridge?"

Shocked, the genie replies, "That's impossible. The ocean is too deep and the distance too great. Is there another wish I could grant you?"

The man says, "Well, there is one thing. Could you give me the key to understanding women?"

After thinking for a long while, the genie asks, "Would you like one lane or two?"

There will come a time when a man is asked to do a woman's job—of selling. Remember, women control 80% of a household's spending. Here's some insight into women.

Friendliness. Introduce yourself, look her in the eye, and shake her hand—not a bone crusher, or too wimpy. This shows you take her seriously and is the start of your business transaction. After all, when you have something to sell that she wants to buy, that puts you in business together.

Eye contact. Don't use a turn-off stare, but a genuine I-see-what-you're-saying kind of interest. Looking someone in the eye implies openness, caring and understanding. It shows that you're paying attention.

Women are most comfortable in buying situations when she feels rapport, a sense of belonging and being understood.

Respect. Show her you respect her time. Ask if she has time so that you can learn more about her needs and what she is looking for. This is a critical win or lose point.

Use the same tactics women use when selling to women. She's not a Sergeant Joe Friday from *Dragnet*. She doesn't just want the facts or statistics. She's looking for honesty and integrity in the people she deals with as well as your products.

She cares about people, feelings, relationships, so the more you can do to link these qualities to your products, the better chance you'll have for a sale, as well as building a long-term relationship with her as a customer.

Listen. Being a good listener means really listening, not just nodding your head. Women love to talk—about people, emotions, relationships. It's not gossip, guys. Women talk about things that matter to them.

It might sound as if she's giving you her life story when she tells you how little time she has to cook because she's trying to get all the kids to their after-school activities in opposite ends of town, and her husband invited her boss home for dinner without sufficient warning, and whenever she cooks for company she always seems to spend her time in the kitchen rather than socializing with her guests, and once the dog jumped up when she wasn't looking and ate half a plate of chocolate brownies and she didn't have time to take him to the vet because her guests were already there.

PATIENCE! Just realize it will take longer to sell to a woman than it will to a man.

Go slow. This goes hand in hand with listening. Take time to find out what she expects from you before you hit her with product information. By going slowly, you'll pick up clues on how to help her in the way she wants to be helped.

Demonstrate that you're actively listening by paraphrasing what she's saying.

Take some deep breaths to slow the pace somewhat.

Involve her in your presentation by asking leading questions. You might ask if she'd like to sample what you have on the grill while you're talking.

Whoa. Unless you want a response of "no," don't overwhelm her with unsolicited information or advice, and save the statistical data for your sports buddies.

Be a gentleman. Be hospitable. Politeness is the key. Find out what she wants and needs before you try to push

Differences between men and women

Women:

- View a sale as a relationship.
- Go shopping
- Talk about people, feelings, relationships.
- More interested in service and relationship.
- Ask a lot of questions.
- Sometimes just need someone to listen.
- Want to be understood and to understand.

Men:

- View a sale as a transaction.
- Go buying.
- Talk about business, money, sports, stats.
- Care more about product and price.
- Think questions are a sign of weakness.
- Want to fix or solve a problem.
- Are competitive communicators. They tell facts, give advice and interrupt more frequently because they want to win or come out on top.

Gender Sell, How to Sell to the Opposite Sex Judith C. Tingley, Ph.D., & Lee E. Robert

whatever you have available. Never call her "Hon," or "Dear." Use her name if possible during your conversation.

Focus. Keep the focus on *her*. Her needs, her family, her concerns. Don't talk about yourself.

Use emotions. Link your products with people and feelings or emotional events. This can be done by inserting a few words that convey a positive feeling into the conversation. Here are a few words that shouldn't make you feel uncomfortable: pleased, enthusiastic, excited, confident.

Avoid negative words such as the following: cautious, uncomfortable, uncertain, bothered.

Another way to inject emotion into the conversation is to refer to key life events, such as a child's graduation, anniversary, or birthday, and relate it to the products.

Soft. Use a soft approach. Recommend rather than command. Avoid using strong words such as "should," "must," "absolutely."

Story marketing. Tell a story. Use examples of how other customers used your products and recipes.

"Does your husband enjoy being a grill master? Do you think he'd enjoy impressing your guests with extra-thick-cut steaks? How about rack of lamb? Our customers rave about how easy it is with these spices we've put together for you. Throw together a salad and some baked potatoes before your guests arrive and you're done."

Flexibility. Since women are more attuned to social nuances than men, change tactics if your customer loses interest or becomes distracted. Switch to a question and answer format to regain her attention. Here are a few suggestions:

Will this just be for you or are you planning something special for guests?

Do you like to experiment with new recipes or stick with the tried and true?

Would you tell me more about your plans?

Above all, show genuine interest...and listen to her.

Guidelines for women selling to men

Turn about is fair play, so ladies, what does it take to sell to men?

First of all, recognize that sometimes you may need to adapt tactics used for selling to men when you are selling to a woman. Be flexible and follow the customer's lead, then adapt your sales approach to their individual styles. Never allow yourself to think your self-esteem is too low to be good at selling. The easy way to boost your confidence is knowledge. The more you know and the longer you share that knowledge, the easier selling becomes.

Know your product. Remember this basic difference —men like to talk about facts and things. Women like to talk about people and feelings. They're more interested in product and price than women, who like to buy things when they're described in the context of people and emotions.

If you haven't already done this, play grass farmer for a week. Follow your grazier spouse around and ask a lot of questions. You want to be prepared for that male customer who will greet you with amused skepticism regarding what makes your products exceptional, and how they got that way.

Take a crash course from your MiG man to learn why paddock shifts are better than open range grazing. Learn how what the French call *terroir,* literally "the ground," and the variety of grasses and legumes (do you know the difference between them?) affect animals' gain and the flavor of the meat and milk, just as different grape varieties grown in different soils make wines taste differently.

What does the breed and genetics you're using have to do with the end product? How can stress ruin a seemingly "perfect" product? Most of all, what makes 100% grassfed products exceptional when compared to grainfed animals?

Confidence. Act confident, even if you don't feel that way, particularly if the man because of his size, age or demeanor appears intimidating. Use eye contact, speak warmly, and realize that true confidence will come easier with experience. If you believe in your product (and you should), then you have nothing to apologize for or defend.

Focus on substance and logic rather than your style or emotions. While men appreciate directness, you can go too far. Don't tell them you're the best, instead tell them how many years of experience you have as a backup to your sales pitch.

Talk. Women like to talk about things that matter, and in great detail. Men view this as gossiping. Don't hit them with everything you know, just be prepared to answer their questions with confidence and knowledge of your product, *when they ask you.*

Don't assume they know everything about your product, which in fact they probably don't. However, don't talk down to them. Offer some subtle education, *when they indicate they'd like to know more.*

Judith Tingley in *Gender Sell, How to Sell to the Opposite Sex,* gives an example of a man who needed to buy a vacuum cleaner because he got new carpeting. The saleswoman explained what a great model he was looking at and asked, "What kind of vacuum cleaner do you presently own?"

The man, thinking "A blue one," walked away saying, "I'm just looking," rather than admit he didn't know anything about vacuum cleaners.

The moral of this story is that a woman's knowledge of her product can lose a sale if she talks about facts and data too soon. "Always find out discreetly what level of product knowledge your customer has and then move forward at his pace," Tingley suggests.

Start a conversation by asking open-ended, low-risk questions that give the guy a chance to reveal just what he does know. Be friendly, nonthreatening as you guide the guys toward a purchase. And give him only one thing to think about at a time.

When the man starts speaking, respond in a manly manner—stand there emotionless, nod your head from time to time, grunt, and don't interrupt.

Guy talk. Remember what guys like to talk about—sports, money, business. If you're not a sports fanatic like your spouse and the only stock you ever bought was a milk cow, at least read the sports and business headlines so you'll know the basic players and teams and current business news.

The Wall Street Journal is great for speed reading, The column "What's New" on the front page gives single paragraph summaries of business and finance and world events. Generally, the center front page column gives an off-beat, sometimes amusing profile of someone with an interesting business.

Developing confidence

1. List every obstacle and objection you can think of and an action or response that would result in a successful sale.
2. Mark a "C" by all the things you can control.
3. Mark an "NC" beside the things you can't control.
4. Focus on the things you can control.
5. List all the possible actions you can take to sway the outcome of a sale.

Gender Sell, How to Sell to the Opposite Sex by Judith C. Tingley, Ph.D., & Lee E. Robert

Inside, the *Journal* regularly features movie and book reviews, food trends, real estate and travel. Try it. You'll like it.

Be a female Sergeant Joe Friday. Guys like the facts, just the facts, ma'am. They want to know the bottom line. Get to the point—in a gracious manner—as soon as the conversation allows. Details can come later if you feel they're needed, or asked for. Talking faster won't help sales. Keep it simple. Clarify when asked to do so. Most guys are too impatient to want to hear a rambling Rose.

The winners, and losers. Women are often facilitators in the conversational arena. Men generally practice one-upmanship. While they may not intend to, men can easily dominate a conversation with a man. The worst thing a saleswoman can do is try to compete with him. Remember, this is not about you and your accomplishments, it's about the *customer*, male or female. If he's telling a story, bow out and let him have the last word.

Good humor gals. Don't be shy about using humor when it's appropriate. In *Gender Sell*, Tingley told the story of two men meeting with two women who were trying to sell their services. At one point in the presentation one of the men finally blurted out, "Are there only...*gals* in your firm?"

One of the *sales*women replied, "Well, as a matter of fact, that is correct. However, I have a Great Dane named Hector, a Siamese cat named Freddie, and a parrot named Bud, and they are all guys. But wouldn't you rather work with us?"

Emotions. Keep your emotions to yourself. Guys get uncomfortable when they think women are getting mushy and emotional. Being logical and rational has more clout when selling to men. The less he knows about you and your personal life, the better...unless he's fishing for a date, but that's another story.

How to make meaningful connections

Spanish proverb: The nightingale will run out of songs before a woman runs out of conversation.

Women love to talk. So let's turn that facility to our marketing advantage by seeking connections with our customers.

Perhaps one reason women talk so much is because we feel it's our responsibility to keep a conversation going.

When you connect, you immediately establish rapport. Prospects will see you as someone like them. This helps put them at ease. It banishes the anxiety that comes from facing a stranger, much less talking with one. Even the most self-assured person will feel a twinge of anxiety when meeting a new person or being thrust into a new situation. Connections create bonds.

Friendliness. It all starts with a smile. Enhance that with eye contact. Be a good listener and include the person's name in the conversation if it can be done naturally.

Jump start connections. Babies make the best connectors. Who can resist cooing over a newborn or cute into-everything toddler.

Lots of people bring their dogs to the farmers' market. You might be able to stop someone who isn't even headed for your booth by commenting on their dog. Keep a jar of dog treats to hand out whether or not anyone makes a purchase. My mailman occasionally brought chicken livers for my dog, who was always excited to hear the mail truck coming down the street.

Comment on something they're wearing, or the outfit in which they've dressed a young child. Sports could be a good topic if they're wearing a T-shirt with a team logo or are escorting a little leaguer on their way to practice.

This could particularly work well with a shy person or someone who's being ignored and who needs an "invitation" to come closer and engage in conversation about your products.

When all else fails, comment on the weather.

Ask questions. Women ask questions to fill the silence. Men think asking questions is prying.

As noted earlier, treat a new prospect as if you're on a date. Use questions to find common ground. Remember the reporter's Five Ws—who, what, when, where, why.

Questions serve a purpose in addition to ferreting out information. They're attention getters. They show you're interested in the other person. By asking about something they're wearing, doing (are they walking a racing bike through the market? pulling kids in a wagon?), or driving (did you notice them arrive in a convertible, or antique car, or some unfamiliar model?) makes them feel important, because you noticed.

Questions allow the other person to talk and can activate the thought process. It can lead to insights into who they are or what they enjoy, which in turn can give you clues to what they need you to sell them.

Make truth statements. Agreeing with someone about something is a subtle way to establish rapport. This is the time to ask questions that have a "yes" or "no" for an answer, such as, "It sure is hot today, isn't it?"

If you make a bad connection with someone who cuts you off cold, don't take it personally. Get over it and try again with someone else.

Keep the ball rolling. Asking one question after another can establish the momentum necessary for a prospect to buy into being a customer. This should be easy because most people enjoy talking about themselves, especially to strangers, and particularly when one question is followed by another.

When you learn something new about a person, make a note later to use in building a customer profile. Write down their children's names, where they attend school, what kind of pet they have, anything they tell you that is worth remembering. Next time you can follow up on what you learned, which makes the person feel special because you remembered them or something important about them.

When a bond is established, asking their opinion of something can add to their importance to you and make them feel valued.

When you establish good relationships with your customers, they'll be more forgiving when something goes wrong. For example, we had ordered a grassfed turkey for the holidays and had to eat beef brisket instead. A pack of dogs (thankfully, not our Basset Hound Rascal this time) had killed the turkeys before they were ready for processing.

TLC for repeats. The best customers will be those who repeatedly buy from you. Treat them special with tender loving care. If they made a purchase from you for a special event, follow up to ask how things went.

Marketplace dialogs

Welcome your customer's input by posting a sign that reads, "How can we better serve you?"

Make a suggestion box or keep a notepad and pen nearby. Have one response already written to encourage more suggestions.

Include a postcard with a question on it with every purchase.

The first week after an initial purchase is a critical time psychologically, and the best time to follow up. Whenever I purchase Chico's clothes out of town and have them shipped home, the package includes a handwritten postcard from the sales person thanking me for shopping with her.

When we bought a new-to-us, used car we received a thank you note with a printed message and a handwritten P.S. thank you. This was particularly surprising since we bought (again) from our best friend's dealership, and they had already verbally thanked us.

Our vet always follows up the day after Rascal receives treatment, like the time they had to drain a huge zit from his neck (five minutes, $15), or the time they had to pop out a beef bone that had gotten stuck between his teeth and upper palate (two seconds, $25), and then there was the time he tried to commit suicide by dining on un-countable number of aspirin from Allan's travel kit, and....anyway, they always call.

Genuine interest. Never feel guilty about spending too long talking to one person. Such genuine bonding can pay big dividends later on. Or not. But it's part of the process of becoming a real person, not just a "take-a-number" prospect.

The exception is that person who has nothing else to do but talk the ears off of corn to the detriment of interaction with other prospects and customers. Prepare ahead of time a phrase to use to politely disengage yourself from such a situation. Say something like this, "I'm so pleased to have met you, but I don't want to take up all of your time. I'm sure you have other shopping you need to do, and I'd better get back to work.."

If it's a repeated occurrence, make an advance plan with someone in your booth (or the one next door if you're alone) to distract them by urgently needing your attention. Now! Be sure to clearly disengage. "Excuse me, but I'm needed elsewhere. It was good to see you again. Have a great weekend."

Remember, people don't listen to what you tell them. They listen to what you ask them.

What makes "feel good" sales?

Studies have shown that when people are in an upbeat mood they're more likely to make a purchase they're satisfied with, because a positive attitude enhances their confidence.

Satisfied customers are those who are more likely to make repeat purchases. And remember, happy customers are best at spreading the word to their girlfriends.

Customers who are in a bad mood are going to be harder to sell. Maybe they're just not feeling well, or have had a fight with their spouse or children. Whatever the cause, they instinctively just don't trust their judgement. They'll unintentionally set up walls of resistance and have that dreaded "no thanks" phrase ready to speak.

Of all the rejection letters I've received as a writer, the only one I remember said, "Editor is having a bad day."

What can you do to make things more pleasant to ease into a sale?

Smile. It takes little effort to present a friendly smile. Be sincere. Cheerfulness can be contagious. Be the first to say hello. Don't anticipate rejection. Instead, look for the positive in others. Smiles signal acceptance and cheeriness and can set the tone for your business relationships. By the way, smiles come more easily to women than to men.

Frowns indicate that something is wrong and could cause customers to wonder if business is bad.

Appearance. Take the extra time to personally look good and make your selling area clean and attractive. This projects an aura of confidence.

Listen. Customers want to be understood. It takes listening skills to acknowledge their concerns and respond to them specifically. They don't necessarily want a new best friend, but it can boost their ego when you remember their name or something about them from the last time you saw them.

I was making a bank deposit one day and the teller noticed the name on the check from my nephew's wife and asked about her. Then she said, "And you're the writer." She had remembered my novel writing from twenty years earlier, even though I don't remember having met her. Don't you feel wonderful when someone tells you, "You've made my day."

Be open-minded, non-judgmental, flexible.

Focus on the customer. Always be thinking about your customer rather than yourself. Her life and her family matter most to her. Ask her how her life is today? What's the biggest impact your products have had on her family? How did what she purchased the previous time work out for her? Knowing personal details is more important than knowing business details.

If you stop working on relationships with your customers, you will lose your edge of intimacy. Money is usually at the root of that issue—wanting more profits and dollars rather than staying focused on your customer's wants and needs. Sometimes just doing a good deed may be the only thing that separates you from your competition.

Marti Barletta in *Marketing to Women* explained that if you meet the expectations of women, you exceed the expectations of men.

Name calling. Introduce yourself. Often this encourages the other person to do the same. Don't over do it, but try to use their name in the course of your conversation. Used too many times, name calling comes across as fake or trying too hard.

Several years ago I found myself in a conversation where actor, Erick Estrada was mentioned. "Oh yes, I like her music," I said, thinking the name was Erica Strada and that she was a singer. All credibility was instantly blown. The moral of this story is if you're unsure of someone's name, don't use a name at all.

Calling someone by the wrong name is the ultimate downer. The whole purpose of saying someone's name is to create a comfortable selling/buying atmosphere.

Experience. Not only should you know everything possible about your products, but you should know your competition's as well. Don't ever put down competitors, which comes across as superiority, but be prepared to explain why your products are better. Talk about the differences between grain-fed versus grassfed rather than naming competitors' names.

Your expertise conveys experience, which helps put customers at ease.

Praise. A little flattery goes a long way. Praise and flattery are second nature for most women. If you don't feel

Creating a positive attitude

"You cannot control what happens to you, but you can control your attitude toward what happens to you, and in that, you will be mastering change rather than allowing it to master you." *Brian Tracy*

"Life reflects your own thoughts back to you." *Napoleon Hill*

"Learn from yesterday, live for today, hope for tomorrow." *Albert Einstein*

"Never mistake knowledge for wisdom. One helps you make a living; the other helps you make a life." *Sandra Carey*

comfortable offering flattering comments, just smile. It's better to do nothing rather than fake cheeriness.

However, the subtle use of flattery can enhance self esteem. Try something like this:

"Not everyone can appreciate the value of eating healthy grassfed products like you."

"You look like someone who appreciates quality."

"You really have done your homework. It's a pleasure talking with someone like you."

Time. It can be a real balancing act to respect the customer's time by not monopolizing their attention longer than necessary and yet spend enough time to make them feel valued.

Let them know how much you've enjoyed talking with them. Tell them you are looking forward to seeing them again. And next time you do see them, make a reference to something they said during the earlier encounter.

Enthusiasm. Be genuine. Let your customers know how much you enjoy being a part of the grassfed movement and helping people make healthy purchases. By making a genuine connection with them you show that their needs matter to you. Be enthusiastic about the interests of your customer.

Missourians David Schafer and Alice Dobbs had folks come up to them and say, "I just had to know what you're selling! You two look so radiant!" As a result they've had vegetarians cross over to eating meat.

Tell a story. While it's important to tell stories or use anecdotes to connect with customers, don't spend too much time talking about you. Remember, *her* needs and desires are what count.

Tell some "you" stories about your customers, not just stories about the products. Share other customers' success stories and testimonials. Target what women enjoy, cherish and take pride in.

Make emotional connections. The main reasons people buy are to solve a problem and to feel better. The words you use can pack emotional power.

Here are some feel good phrases: Highest quality, top of the line, guaranteed tender.

Paint your own power images—talk about your contented animals, grazed all their happy lives in lush green pastures.

Be enterprising. A bit of creativity can overcome sticker shock for new customers. Help customers stretch their food dollars by suggesting lower priced alternatives with recipes that can make meals just as special as those using pricier cuts. Be prepared with low range, mid priced, and high end suggestions.

Jill Bamberg, who wrote *Getting to Scale, Growing Your Business Without Selling Out*, points out that the only skill a business must have is marketing skill. "Creating customers is *the* business of the business."

What a pleasure it is to have happy customers.

Multitasking marketing

We're wives, mothers, girlfriends, marketers, housekeepers, bookkeepers, chauffeurs, and...you fill in the blank. Multitasker should be every woman's middle name.

If we're busy juggling all the varied aspects of our lives, our (female) customers are equally preoccupied. Anything we can do to save our customers time and stress is worthy of their attention, and more likely to lead to a sale.

Let's start with your basic grassfed beef products—steaks, roasts, brisket, and lots of ground meat. How can you turn that purchase into a time saver?

Recipes. Take the thinking out of the decision making for your customers.

Offer a Featured Recipe of the Day with a printed sheet with easy steps for preparation and suggestions of what to serve as accompaniment. Celebrate the pleasure of home cooking and the family dining together.

Turn each recipe into the main course for a special event: *tenderloin* for tender hearts—a Valentine's Dinner for two; marinated and slow-cooked *roast* for Mother-in-law Visiting; quick and easy *ground beef* sluggers for After Baseball Practice.

Give your recipe suggestions quirky names and whimsical reasons to gain attention and show empathy for other women's busy lives: The Grandkids are Coming; I'm on a Diet

Dinner; Candles & Music; Let's Have a Party; I Don't Want to Cook Tonight; First Date Home-Cooked Meal; Burgers to Bring the Kids Back to the Table for Dinner; or simply a nice juicy steak To Show His Grilling Skill with Fire and Knives.

If you're selling at a farmers' market, use a chalkboard to suggest "sides" based on what other vendors have to sell that day, such as fresh grilled asparagus or just-picked lettuces and tomatoes with basil plucked from another booth's potted herbs.

Pre-package meals. Take the recipe suggestion one step farther. Include some herbs or vegetable complement from your farm. Laura Ritch from Goose Pond Farm in Hartselle, Alabama, makes up spice jars of "Laura's Herbs de Provence" to go with lamb sales.

For those of you who like to crochet, offer colorful, old-fashioned egg cozies to everyone who buys a dozen eggs. The cozies could become collectible incentives to encourage repeat purchases. Use a different color pattern for each month of the year.

Sankow's Beaver Brook Farm from Lyme, Connecticut, has a professional cook (most state laws require cooked food to be prepared in commercial kitchens, so check regulations) ready-to-heat lamb Shepherd's Pie, lamb sausage, and lamb stew. They also sell cheese from their Jersey cows and sheep, and prepared *pesto*.

One-stop shopping. Those of you who have multiple products to sell are in an ideal position to take the guesswork out of meal planning. For example, Jonathan White of Bobolink Dairy sells artisan bread with his artisanal cheeses. (Okay, he's a guy, but obviously one who understands his multitasking customers, plus he has women doing the main selling in his booth.)

Ryan and Kristine Jepsen of Grass Run Farm in Dorchester, Iowa, save time for their customers by using their farmers' market booth as a delivery site for pre-orders.

Catalogs. Catalog shopping is one of the biggest time and energy savers for your customers. Think of the time and fuel it takes just searching for a parking place in a crowded lot. Send out a periodic "catalog" newsletter or email lists of what you have to sell. Take pre-orders from your farmers' market booth.

David Schafer and Alice Dobbs in Missouri went to the Kansas City Farmers' Market once or twice a month and only for a couple of hours. Nearly everything was pre-sold. Knowing this, their customers made a point to show up before ten a.m. or else they'd find something they wanted already sold. They always announced the next sale date and had a pre-order sign up sheet available.

Delivery. Joel Salatin delivers to his Metro-Buying-Club ladies on Wednesdays, restaurant sales deliveries on Thursdays. Everything is pre-sold from phone and email orders. He makes one drop at a central location for each of more than a dozen Buying Club groups. Don't forget to add a delivery charge, after all, catalogs add fees for shipping and handling.

Wendy Taggart of Burgundy Beef in Grandview, Texas, makes two weekly deliveries in the Dallas Fort Worth Metro area. This not only saves her customers time, but it gives Wendy a chance to interact with customers to learn their interests and concerns. "The primary key to success," she said, "is word of mouth. Unless we have direct contact with our customers we don't know which cuts they like and why, or their cooking limitations. It's a good way to test market. Even bad

Multitasking

According to *The Merriam-Webster Dictionary*, the term "multitasking" has only been in use since about 1966. That was "my" era, when women flooded the workforce. It was the era of women's lib, a time when women claimed to "have it all," both a family and a career.

It was about time our multitasking talents were recognized.

feedback lets us know quickly what to change."

Joe and Julie Morris of Morris Grassfed Beef in San Juan Bautista, California, sell their beeves two ways. One is a split half priced by the pound and the other is "CSA style" where the customer buys a split half but the Morris' store it and deliver it in three deliveries over the course of the year. This beef is priced about 80 cents higher per pound plus a $20 delivery charge. But this allows customers who don't have the freezer space to still enjoy their beef.

Gift certificates. Encourage your customers to help you get new customers. Be creative in suggesting reasons why they should buy a Gift Certificate— for newlyweds, for a couple with a new baby, to welcome a neighborhood newcomer.

Create a Beef of the Month Club to encourage repeat sales and feature a different cut each month. This could be an opportunity for you to really get creative or simply to test unfamiliar cuts such as French-cut steaks or flatiron steaks.

Remember, the more you can do to save your customers' time, the more likely they'll be to return time after time.

Helping stressed out shoppers

Women love luxuries, right? However, the best luxury of all is not a fur coat, foreign car, or diamond jewelry. It's time.

Women are time starved, says Marti Barletta, author of *Marketing to Women, How to Increase Your Share of the World's Largest Market.* To research her book, Barletta conducted focus groups and learned that women feel pressure from their husbands, from their children, from their work. As one woman put it, "Some of us don't even get to go to the bathroom by ourselves."

So, what does that have to do with direct marketing?

If you want these time-crunched women to buy, you have to make things easy for them.

First of all, you need to understand a little psychology about how women buy. It's almost a contradiction that these time-starved women will actually take longer than a man to make a purchase. Women want time to comparison shop, time to think about the potential purchase, time to do some research. This really throws sales*men* who are accustomed to the more direct male approach to buying. Men generally know what they want to buy before they shop for it.

"Women go through a deliberate decision process before they shop," Barletta explained. "When she is ready to buy, she actually pulls the trigger pretty quickly."

If she says, "I'm just looking." That's exactly what she's doing. Back off. Give her time. But be ready to help when she's ready.

Mary Lou Quinlan, author of *Just Ask a Woman, Cracking the Code of What Women Want and How They Buy*, said that women actually enjoy shopping when they're not stressed for time. She estimates that's about two weeks in a lifetime.

How can you help stressed out shoppers?

Respect. For women, waiting time is the same as wasting time. Don't keep her waiting. End that cell phone call, politely disengage from a conversation with a friend (or customer who has completed a purchase), and give her your full attention. At the very least acknowledge her presence, "Hello, I'll just be a moment longer." That's a sign of service.

"Businesses that recognize and use more women's talents will be more innovative, productive, and more profitable," noted Judy Rosener, from the Graduate School of Management, University of California at Irvine.

Guerrilla marketer, Jay Conrad Levinson, said that

Shopping 101

- Women develop bonds by talking. Men bond by doing things together—playing ball, fishing, going to the big game.
- Women rarely buy early in the decision making process. Men are ready to buy before they shop.
- Never mind how long it takes to make that first sale to a woman. The bonding that develops between you means far more than a single sale. Women will tell their girlfriends about a good experience, and the payoff will be repeat sales and many referrals.

Marti Barletta, *Marketing to Women, How to Increase Your Share of the World's Largest Market*

businesses that use strategies to save their customers time grow three times faster and profit five times more than competitors who ignore the crucial need for speed.

Interaction. Be available to answer questions she might have. Feedback helps her understand the things that set your products apart. She will walk away if she feels she's being manipulated or talked down to.

Convenience. Can you do anything to speed things up? Offer pre-bundled products—ready-to-grill hamburgers, lamb chops with a bottle of spices. In today's world of convenience foods, many women have never learned or have forgotten how to cook. Take the guesswork out of it for them. Anything you can offer that's pre-packaged, ready-to-serve will solve the problem of what-to-eat-tonight.

Leighann and David Wright in Alexandria, Alabama, added a drive-up window to their on-farm store.

Do you accept credit cards?

Go the extra mile. Here's a really off the wall idea that worked. During special promotional periods, Steinway & Sons in New York stages concerts with a grand piano in a customer's home. It results in sales to one or two friends of the customer who were in the audience. Okay, so we're not selling high ticket pianos, but we can do our selling in a customer's home by cooking a special presentation dinner. In-home chefs are replacing ordinary caterers in urban areas. They do the shopping and cook the meal. All the customer has to do is provide the guests. Borrowing an idea from the makeup industry or Tupperware parties, how about offering a downhome version of a "Hamburger Helper" night of demonstrations on various hamburger options?

Extra help. If your farmers' market is a large one, are you set up near the parking lot to save your customers time in finding you? Can you employ one of your children to carry bags for your customer?

I ran into a neighbor getting out of her mini-van with her two pre-schoolers. "I can't get anything done when they're with me," she said. I was in and out of the store long before she reached the check-out line.

Maybe you could help set up a safe child-care area at the farmers' market to give young moms some time to shop. If you helped contribute to the cost of toys, set up, staffing, make sure your farm's name is mentioned as a sponsor.

Understanding. Find out what results she is looking for. Ask how, where and why she is shopping to better understand her needs.

"The sweetest words you can tell your stressed out customer," said Marti Barletta, "is 'I'll find a way to make this work for you.'"

Overcome price objections. If she hesitates at the price, borrow a line from the cosmetics industry. Remember the advertisement for hair care products that says, "Because you're worth it." Or how about, "Live a little." After all, what is life without a few luxuries?

Stressed out sellers

A candle loses nothing it if it is used to light another. Unfortunately, most of us are burning the candle at both ends. That can lead to burnout.

Let's talk about you.

Perhaps you're feeling pressure from family obligations, pressure to take on the direct marketing when you have no experience at selling, or even pressure to lose weight, gain weight, or stick to a regular exercise routine. The source doesn't matter. When you're stressed, your emotions are on edge, and your performance suffers. Mistakes are more likely to occur when you're exhausted. Making simple decisions can lead to agonizing second guessing—did you make the right decision? Who knows. More importantly, who cares?

Steve Jobs, founder of Apple computers asks himself every morning, "If today were the last day of my life, would I want to do what I am about to do today? And whenever the answer has been 'no' for too many days in a row, I know I need to change something."

Taking care of yourself is your first obligation. If you're not in good physical and mental shape you can't perform at your optimum level. That doesn't mean there won't be times when you go on autopilot just to get through the day. However, too many days like that aren't good for you or anyone around you. Sometimes you have to set aside time for yourself.

Whenever we drop our dog, Rascal, at the kennel, I always pay extra for "Me Time." That means someone spends 30 minutes a day brushing him, playing with him, feeding him treats. When he's in a bad mood he doesn't get Me Time. He's left alone.

Here are some suggestions for your own Me Time.

Find your own balance. Don't make comparisons to anyone else. Find your own definition and means of balance. Comparisons will only lead to guilt trips or make you feel inadequate. You are uniquely you, and there's nothing wrong with that.

Niki Joy, author of *Selling is a Woman's Game*, suggests looking at everything in regard to the three Cs: commence, continue, conclude.

Be flexible. Realize that as the business grows your balance can shift. There will be times when you have to put family before business, and times when business comes first. Don't let anyone denigrate you for putting one over the other when special circumstances dictate your choices.

Have a plan for your day, but give yourself permission to change the plan when necessary. Make a habit of living one day at a time, and enjoying the process.

Me time. Some of us (me) need eight or more hours to

The five rules of happiness

1. Free your heart from hatred.
2. Free your mind from worries.
3. Live simply.
4. Give more.
5. Expect less.

Herman Cain said, "Success is not the key to happiness. Happiness is the key to success. If you love what you are doing, you will be successful."

feel fully rested, others need less. Know what works for you and then make it a priority to get a good night's rest. Take naps, even if it's only for 15 minutes to recharge. Don't take problems to sleep with you. Try to relax with a blank mind.

Set aside time for yourself, even if it means locking yourself in the bathroom. Meditate, pray, veg out, read a book or exercise.

Hide from your kids. Drive somewhere and just sit in the car with the seat reclined and the windows open to catch the breeze.

Take a pasture walk. Alone.

In that quiet place review your achievements and how good that made you feel. Go over in your mind compliments people gave you.

Write down your achievements from the last few months and periodically go over these with your spouse to see how far you've come. Consider this a record of your growth curve.

Not only should you allow yourself down time, you should schedule it. And don't use that time to catch up on tasks left undone.

Our time. Set aside time to be alone with your spouse. Have a regular date night, even if it's just once a month. Set aside time for your family, perhaps during meals with no TV allowed.

Celebrate the pleasure of home cooking with the whole family involved. Rod Creher, author of *Crunchy Cons* said, "The rituals surrounding food preparation not only nourishes the body, but it nourishes something in the human soul."

Exercise. Build recreation or exercise into your weekly schedule. This may mean getting up before everyone else or being the last one to turn out the light. The important thing is to be consistent, every day.

Organize. Whenever we drive into town (about 20 minutes away), we make a grand loop including everything that needs to be done at various locations. This not only saves fuel, it saves time and energy. Prioritize your errands in case time runs out and something has to be left for another day and trip.

Often I feel stressed because I'm juggling tasks and things to do in my mind, going over the same things so that I won't forget them. When I write things down and prioritize, I realize the list is much shorter and more manageable than I imagined. (Don't you love having a list so that you can cross things off of it?)

Use one calendar. Expert jugglers use one big calendar with enough space to keep track of your multiple roles—shopping lists, family members' appointments, calls to make, important events, points to remember, things to think about. Use a different color for each family member.

Experts believe habits form in 18-21 days, so commit to trying the calendar system for one month.

Friends time. If you can't go to or afford a spa day, create your own. Host a girls' night out with a health and beauty consultant, scrapbooking group, or home decorating party.

When you take on new or extra tasks, time formerly used for seeing friends or family can disappear. You can maximize your social life. Plan group gatherings—a family barbecue, lunch with all of your girlfriends rather than one-on-one get togethers. Stay connected.

Putting senior customers first

As I write this, my mother is 97 "and a half" as she'll add when asked. She is living independently in her own home. However, recently the time came to introduce a part time aide into her weekly routine.

Following leads friends and family had given me, I started interviewing.

Candidate number one told me, "I need to come in the morning." I explained that mother makes her own bed, dresses herself, takes her pills, fixes a cereal-and-milk breakfast and a sandwich for lunch. She needed someone in the early evening to prepare a hot evening meal, spend some time visiting, and stand by while she showered and got ready for bed. "That's too late for me," number one responded. "What if I come between one and four?" That wouldn't help mother or me since those are her napping hours.

Candidate number two was having a bad day and told me about it. Her car wouldn't start, and she didn't recognize the name of the person who had referred me to her. Although she verbally agreed that keeping an elderly person's routine as disruptive as possible, she gave me the impression that her service would involve a parade of people—whoever was available on her schedule. "I'm not sure which of my girls I'll get for you, but next week it will probably be me," she said.

Number three was an agency of caregivers who clearly

understood the importance of putting the customer first. When could they come? "Whenever you need us." What tasks would they do? "Whatever you need us to do."

What does this have to do with marketing grassfed products?

Stocking healthy, nutritious meals in mother's refrigerator has become a challenge for me. I only cook once a week.

Who invented portion sizes anyway?

That's what a 5'10" 170 pound martial arts instructor asked on the everydayhealth.com forum.

"If I measured out a single serving of everything for dinner—meat, potatoes and vegetables— it wouldn't feed my 10-year-old," the instructor complained. "Is that a serving for a 5' 90 pound woman or a 6' 250 pound active man?"

The Portion Plate™ is a melamine plate illustrated with items to reflect recommended portion sizes for a daily meal. A tennis ball represents a piece of fruit. A deck of cards represents a portion of meat. The kids plate uses a box of 8 crayons in place of the card deck. Half of the plate should contain fruits and vegetables.

Here are some portion cues from checkyourhealth.org:

Handful = 1-2 ounces.
Fist = 1 cup.
Palm = 3 ounces.
Thumb = 1 ounce.

Who invented portion sizes? Even Google doesn't know.

That means I can assemble one, perhaps two meals from Sunday's leftovers for mother to take home. Due to her dietary restrictions, I don't like to buy take-out meals since I can't control the ingredients—too much sugar? Salt? Carbohydrates? Spices? While there is a wonderful local business that sells casseroles, they are frozen, and entirely too large for mother's simple needs.

This got me to thinking about grassfed options for seniors. Many in their upper decades just don't eat the hale and hearty meals they did when they were younger. Many cut back on meat. Many are widows or widowers. However, those who loved a juicy steak when they were younger, would enjoy it again if they could find one that matched their needs. They still care about healthy, wholesome meals.

Why not cater to those mature ladies and gentlemen who still want excellent food, just not much of it?

Tender loving care. My mother has never lost a tooth. That's not the case with others who grew up prior to the age of preventive dentistry. Meat tenderness is a vital element for seniors. They're likely to shy away from meat simply because they think it might be tough. And don't forget price consciousness.

Since one tenderloin steak might provide up to three meals for the mature meat lover, why not package up less expensive tenderloin tails in mini quantities? Three small, individually sealed packets could be combined for one value package. You could do the same with one ribeye steak, one New York strip, or even a flatiron steak sliced into smaller portions. Rather than expect the customer to cut up a steak, bag each portion, and place several bags into a larger one for freezing, pre-wrap the smaller portions for them. It will require more cut-and-wrap on your end, however, remember it will benefit the customer.

Be a pioneer. Before the days of refrigeration, the meat people ate came from small animals—rabbits, birds, squirrels. Unless it was a special occasion with a large gathering of people to share the meal, the meat from a larger animal would spoil before it could be eaten.

It's difficult to separate individual lamb chops, which are already small by nature, without thawing the entire package. Even if you manage to cut out two from a group of four, the original wrap is compromised. Instead provide two chops sealed together rather than four.

Those of you who are already selling hamburger patties, give your customers the option of purchasing smaller rounds—call them Medallion Burgers or Burger Nuggets. Then simply group those smaller, individually wrapped items into a larger package so that price-wise you'll come out about the same or maybe even a little higher due to the extra wrap and time.

By the way, did you know what made drugstore soda fountains so "cool"? Home ice boxes were either too small or too inefficient to keep ice cream solid.

Think small. Seniors aren't the only ones who will benefit from smaller portion sizes. Mini meat packages can encourage children to try something new without the waste of a full size regular cut of meat. Mini meats can add diversity to meal planning. Smaller cuts can beef up breakfast. Save your customers time cutting up a larger piece of meat for appetizers or party food. Do the work for them.

For those of your customers who are on a diet, smaller portions will likely meet their daily requirement for meat. In fact, you might promote them as Diet-Sized portions.

Anything you can do to downsize portions will make you a pioneer in catering to the largely unaddressed senior market. I'm guessing it will only be a small part of your regular business and could very well be extra trouble for you in the extra packaging required. However, those customers who are caring for seniors will love it. You'd certainly get a hug from me, as well as plenty of new customers who are friends with aging parents.

Multi meals. Mix and match smaller items into one overall package. Make up Meals in Minutes, or The Fabulous

Four—a beef, burger, lamb chops, and kebabs—enough to provide one meal for four days. Keep each item individually wrapped.

Those of you who are adding value to your products by cooking batches of stews, soups or roasts ready for reheating, repackage these into smaller portions.

Shout it out. We all know that being hard of hearing tends to go with aging. Be sure your customers hear about this. Make a big deal over this new line. Give the mini meals catchy names. Write press releases about it. Promote it at your farm store or farmers' market. Remember, anything that's new or different or a change from what you normally do is worth calling attention to it.

Theatre: making food fun

Shakespeare knew it—"All the world's a stage."

The simple act of someone watching what another person does is theatre.

Often that action takes place off stage, such as in a restaurant where meal preparation is hidden from the diners.

Increasingly, smart restauranteurs have put cooking center stage by designing open counters that allow customers to watch chefs at work sizzling steaks or burgers on the grill, flaming desserts, tossing pizza dough in the air, or rapidly chopping sushi up close—too close?—and personal table-side.

Like magicians who reveal the "secrets" behind their illusions, food marketers who understand the benefits of theatre know that making food fun creates a memorable experience by engaging the customer's senses, and encourages repeat performances.

Remember your audience. Always be aware that your audience—the customer—is watching.

Consider this scenario: A checkout clerk is chatting with a bagger about what they're going to do when they get off work while the checkout clerk is going through the motions of scanning your items. Would you call their actions customer-focused? Is there any customer service implied when your total is announced, and robot-like you are asked for your credit card?

Compare that experience with this one: Barbara Glanz

tells the story of Johnny the Bagger, in *The Simple Truths of Service* (written with Ken Blanchard). At a presentation on customer service to a supermarket chain, she challenged each employee to create a meaningful memory for their customers.

Excited by the idea, Johnny, a bagger and a Down syndrome youngster, spent several nights looking for thoughts for the day. Sometimes he made them up. His father helped him enter into the computer and duplicate his list. Johnny cut out and signed each quote. At the market, he dropped one thought for the day in each shopper's bag and said, "Thanks for shopping with us."

In no time, people formed long lines at Johnny's checkout lane. The manager tried to open other checkout lanes, but everyone wanted one of Johnny's quotes.

Johnny's action inspired the floral department to turn broken flowers into corsages to give to elderly women or little girls. The butcher customized his meat wrappings with Snoopy stickers, reflecting his love of the cartoon beagle. All of these actions, simple in themselves, got people's attention. They told their friends about the store, made repeat purchases, and as a result increased the bottom line.

Remember you're always on stage. No action goes unnoticed. Even when you're not making a presentation to a customer, be aware that you're "on stage" in the farmers' market booth, at your farm store, and on the farm itself. Your actions should reflect your professionalism. Don't be putting on your makeup where a customer might be watching. Curb cell phone conversations unless they are directly related to your customers. Stay focused on the business at hand.

Building stage sets. Everything you do makes an impression on customers. You can create a sense of flair and originality by the way you arrange merchandise on a shelf. The merchandise itself can send subtle messages.

At Burgundy Pasture Beef in Texas the Taggarts sell rock salt lamps. The heat from a light in a salt lamp emits negative ions, which are considered as healthy as a trip to the sea or mountains. It's another sign of commitment to the health of their customers.

Women love flowers. Incorporate them into your displays. These can be unique using wildflowers from the farm or blooms from your own garden.

Assemble gift baskets to take the guess work out of what to serve with your core grassfed products. Not only can this solve gift-giving problems for your customers, but it can inspire and direct them toward a particular purchase.

Marlene Groves sells buffalo meat in the Denver farmers' market, which is already heavy with other buffalo vendors. To stand out from the competition she has a hide instead of a tablecloth on her table. Not only customers but other market vendors comment on how nice her display is.

Enter stage left. When I was a child I was always organizing my friends to put on a play. Invite your children to write a skit and perform it for your customers. This could be a form of "edutainment," a cross between education and entertainment, about some aspect of farm life. Let them make costumes, create a talking Grassy the Cow. After all, the California Cheese Board commercials say happy cows make happy cheese. Keep it fresh, keep it fun.

Do you or your children sing or play a musical instrument? How about "staging" concerts once or twice during the course of a market day, or making this a bigger production on the farm once a year? This may be an offbeat idea, but it does

Keep your products center stage

"Product preference is perishable, especially these days. Unless you come up with new ways of recreating or reestablishing that preference, of reselling people on your product, they will disappear—and so will you."

Sergio Zyman *The End of Marketing as We Know It*

attract attention. Mix it up. Keep it surprising. Make shoppers want to hang out at your place. Give them another reason to leave with a smile on their faces.

When a customer purchases salmon in the open air World Famous Pike Place Fish Market in Seattle, one fish seller shouts out, "One salmon flying away to Minnesota," and then tosses it several feet in the air to a colleague who catches it, wraps it and takes the customer's money. Owner John Yokoyama, quoted in *Authenticity*, said, "Our secret to success lies in our commitment to *being* who we say we are. *Just be it.* Your challenge is to 'just be' who you want to be."

Lambert's Café calls itself "The home of the Throwed Rolls." This restaurant chain offers home-style cooking and plenty of laughter. After ordering an entree, waiters constantly pass through the restaurant with "passalongs," buckets of side dishes—turnip greens, mashed potatoes, fried okra—and they toss hot rolls clear across the restaurant to customers. John Phillip Sousa-style marching music keeps people ordering and eating, and coming back for more. It's fun. It's theatre. It's memorable.

The Flack Family Farm in Enosburg Falls, Vermont, stages the Raw Milk Theater variety show once each summer in the evening. They invite friends and family to entertain about rural life, the history of farming, and "the affairs of milk, food and good health." It's like an old-fashioned vaudeville show where performers act in front of a canvas with a painted rural scene. And surprise, it's free. The audience is asked to bring their own chairs, family and friends.

Y'all come back. In their book, *The Experience Economy, Work Is Theatre & Every Business a Stage,* B. Joseph Pine II and James H. Gilmore have updated Shakespeare's observation and applied it to businesses today. They suggest you make the place where you sell a "trade show" experience. They describe this as "a place to find, learn about, and, if a need is met, purchase offerings."

Pine and Gilmore wrote, "Instead of saying, 'Have a

nice day,' which characterizes most unmemorable service interactions, ...use an expression befitting a memorable event: 'Did you have fun today?'"

When you make the shopping experience memorable, your customers will not only return for more, but they'll spread the buzz to their friends.

Luring customers with samples

One Saturday we stopped at a roadside fruit and vegetable stand. Twice.

We weren't the only ones who made purchases at that stand twice in the same day.

Although it wasn't on my "want to buy" list, we bought a basket of peaches because the farmer offered sliced samples. (He also had samples of watermelon and cantaloupe, but we had previously bought each of those elsewhere.) At home, Allan and I finished off most of the oh-so-sweet peaches before lunch, so we *had to* return for more that afternoon.

While Allan was buying two baskets of peaches, a woman who had purchased a cantaloupe that morning drove up. "You'd better get one," she advised, laying down money for more melons. "They're really great." Noticing behind the farmer that an open-air bin the size of a commercial dumpster that had been full of cantaloupes that morning was then nearly empty, we bought one too.

We went back again the next weekend.

Samples can tip the scale when it comes to sales. This is particularly true with meat and cheese products. What can be more enticing than the aroma of grilling meat? The memory of tasting a juicy and tender grassfed beef kebab or a slice of melt-in-your-mouth cheese is an experience that can have lasting effect.

Evan Parry, a graduate of the California Culinary Academy keeps a grill going in Fox Fire Farm's booths in the Durango and Telluride Farmers' Markets. He admits that often people duck into the booth for a sample, hoping to escape without a sales pitch, but once they sample the Colorado farm's grassfed lamb, they're hooked.

"Most will come back and talk to us about our lamb,"

Charge for experiences, give away your products

As grassfed meats become more popular, entering the mainstream, they will eventually become commodities.

How do you separate what you sell from everyone else?

Authors Joseph Pine and James Gilmore of *The Exprerience Economy* suggest giving away your products, while charging customers for an experience.

How?

Charge for admission to field days or farm tours and include a free lunch. Charge for parking and use that fee as "credit" for purchasing meat, milk, eggs.

Greg Judy plans to "invite" his neighbors (free labor) to participate the next time he has to trail his herd over county lanes from one farm to another. In exchange he'll host a (free) cookout of grassfed meat, solidifying community relations. One day he may even charge anyone who wants to help.

Think of ways an experience can engage your customers and give them lasting memories that they'll want to share with friends.

he said. "If you can get them talking, you've got them."

Several years ago, Allan and I attended the Ag Expo in Paris. In addition to exhibits of various breeds of live animals as you would see in USA stock shows, breeders passed out samples of cooked meat from each breed. Apparently it was a requirement for exhibiting. That tactic really demonstrated the diversity of flavors from breeds raised throughout France. The *terroir*, or "taste of the land" made the experience akin to a wine tasting—with beef. Flavor comes from a combination of exercise, sunlight, soils and greens.

Oversupply. In response to the question of how to sell their excess ground beef, Wendy and Jon Taggart of Burgundy Beef started selling cooked gourmet hamburgers in their Burgundy Boucherie shop.

Initially, they were concerned that a $4.99 one-third-pound burger was somewhat pricey for the store's location in tiny Grandview, Texas. However, the day they opened they sold over 100 burgers with nothing but a sign outside for advertising.

Not long afterward when the *Fort Worth Star Telegram* newspaper did a food-section feature on them, they not only sold cooked hamburgers, but practically sold out of everything they had.

"A lot of people who came down here for the first time were skeptical about grassfed beef and wanted to try some before buying it," Wendy recalled. "So we got the idea of putting in a small kitchen and serving a ready-to-eat hamburger as a way of convincing people to buy our meat."

"People who drive down here from Dallas or Fort Worth to eat a hamburger are ready to spend money," Jon said.

Sure, it's more trouble to haul out a grill, set it up and clean it up afterward (especially when it's still hot). However, offering samples can set your products apart from the competition.

"Selling is *helping* people buy something that they might not have bought without your assistance," note Nicki Joy and Susan Kane-Benson in *Selling is a Woman's Game, 15 powerful reasons why women can outsell men.*

Be aware that demonstrations alone won't sell anything any more than stand-alone facts will. You have to get people involved. Use samples to reel in new customers.

Dare to compare. Joel Salatin used to make egg comparisons, cracking a store-bought, watery-looking egg next to one of his hard-shelled, firm, orange-yolk farm eggs. He used lab tests to show the astonishing cleanliness of his farm-raised chickens compared to the typical supermarket hen.

How about comparing the cloying, grease from grain-fed hamburger to a grassfed burger with its healthy, unsaturated fat? Burgers don't have to be lean to be healthy when they're grassfed.

Why go to the trouble of making comparisons? Women are comparison shoppers and generally won't buy until they've covered all the bases. Whenever you can give her the complete picture by doing the comparisons for her, you speed up the decision making and buying process and earn her trust. Even a simple chart with comparison figures can make a big difference.

Pricing. If you still end up with way too much of something, should you offer it at a lower price? No! Instead, turn that excess into cooked samples that can potentially draw in new customers.

Reading between the lines to close the sale

Anthropologists note that 90% of emotional communication is non verbal. Women and men not only think and communicate differently, but women are better at reading unwritten messages, or body language.

Females pay more attention to body cues—voice tone, hand gestures, posture—while men are more likely to tune in solely to words.

Perhaps one of the most obvious body cues most people recognize is arms tightly crossed against the chest. It could mean the person is cold, but more likely it indicates extreme reserve, a skepticism or reluctance to accept whatever message is being presented.

Arms relaxed at the side indicates an openness. This is someone who is interested in what you have to say.

Are you one of those people, like me, who says, "No," when you don't necessarily mean "No"? For me, "No" means I need more time to think about the question before making a decision. For example, Allan might ask, "Do you want me to grill for lunch?"

"No," I immediately respond. No, really means, "I don't know." I'm busy working on something else now. I haven't even given lunch a thought. But grilling means he'll do most of the cooking and I'll only have to throw a salad and a few veggies together. "Yes, great idea. Let's grill."

When you're selling, the worst thing you can ask is, "May I help you?" It's the kiss of death, because it comes across as being pushy.

How many times have you responded, "I'm just looking." What you're really saying is, "I want to be left alone. When or if I'm ready to buy something or need your help I'll ask for it. Meanwhile, I need some time alone to make up my mind."

It's far better to engage a prospect in chitchat. Comment on something they're wearing, the weather, anything that's not about your products. This establishes a connection and openness to draw the other person in. It slows things down and allows the prospect time to figure out what they want.

When someone gives you a vague answer for what

What women say and what it means

Okay men, pay attention. Here are three things women say and what they mean:

FINE. This is the word we use at the end of any argument when we feel we are right and need you to shut up. NEVER use FINE to describe how a woman looks. It will cause you to have one of those arguments.

FIVE MINUTES. This is half an hour. It is equivalent to the five minutes that your football game is going to last before you take out the trash, so women feel it's an even trade.

NOTHING. This means something. Be on your toes. NOTHING usually describes the feeling a woman has of wanting to turn you inside out, upside down, and backwards. NOTHING usually signifies an argument that will last FIVE MINUTES and end with the word FINE.

Posting from the Internet

they're looking for, they generally are uncertain of what they want. As a shopper, I defuse confrontations by saying, "I'm not looking for anything specific, just something that jumps out and says, 'Take me home.' I'll know it when I see it." This is a situation where you might gently steer the prospect toward a sale.

In *Selling is a Woman's Game*, Nicki Joy said most men don't have a clue about non verbal messages. "Whereas women have a full-blown antenna, men seem only to have a coat hanger with a little piece of aluminum foil wrapped around it."

Here are body language clues Nicki Joy identified to help you understand what the prospect is really saying, and quite possibly may not even realize themselves.

Asking questions. Men think that asking questions is prying. Women instinctively know this is a sign of interest and involvement. Respond by giving informed replies.

Lean in. A different sign of interest that can lead to a decision. Speak warmly. Be genuine and enthusiastic about your subject.

Moving back. Putting distance between you could indicate they're ready to leave, or maybe they just want to slow down the decision making process. Try this trick—speak softly to draw them back in so that they have to come closer to hear. It's a subtle power play on your part.

Repeating questions. This sign of interest means they need more information. Offer a few facts to help them fully understand the product.

Men clear their throats. Women moisten their lips. This is a sign of nervousness, a signal to defuse the situation. Back up and briefly change the subject.

Twirling their hair. When women start playing with their hair they're nervous and need reassurance that they're making the correct decision.

Touching. Someone who touches the sample product or reaches for a pen or the order form is ready to complete the sale.

Nodding "yes." Ah, you've nearly sold them. Reinforce their acceptance by saying something like, "So you agree that....(fill in the blank)....the roast, then would work best for your Sunday menu...don't you agree?" (Use only one statement of agreement.)

Ownership statements. They begin speaking as if they've already made a purchase, such as this, "That would be the perfect thing to serve for our open house." Slow down. Acknowledge their involvement with a similar statement of your own, then close the sale with a question. "Of course, the roast will serve a lot of people. Will one be enough, or do you need two?"

Speedy talking. When they start talking faster it can mean they're ready to buy, or are enthusiastic about your products. Follow their lead and their pace with your own enthusiasm and summarize the key points for the sale.

Slow speaking. They need a slower approach and more time to think about the decision. Give them space and time by stepping or leaning back. Offer suggestions for them to process. "Is there anything else you'd like to know about....how we raise our animals....what we do to guarantee tenderness...what other customers have said?"

Speaking stops, listening begins. This could mean they're wary of your message or still thinking about it. Take this opportunity to give them more details to consider. Confirm that the prospect has all the information they need. Summarize what they just said. Reassure them with new information or repetition of something they said. And close the sale.

Whoa. Some people fiddle with their ear lobe or put a hand over their mouth as a subconscious interruption signal. It's similar to a school child who raises their hand to get the teacher's attention. Whatever you've been saying, stop, and ask, "How does this sound so far?"

To make a sale, capitalize on your ability to read the unspoken messages and tell your customers what they want to hear.

Closing. Once a woman is engaged by your message, you can move to close the sale. Since a certain amount of assertiveness is required, you may feel nervous in the beginning.

Don't worry. Being assertive is not nearly as important as how you position yourself to close the sale. Here are some things you can practice doing until selling and closing that sale becomes second nature:

Repeat. If you've been a good listener—and most women are—you can repeat the top three concerns your customer mentioned to you. Then bridge those ideas with three benefits your products offer as answers to her needs. Basically, you're just summarizing what was previously said while keeping the customer's needs in mind.

Advice. Share your expertise, not only on your products and how you personally have used them, but giving examples that other customers have shared with you.

Race against time. If she still needs a nudge to take action, put the sale in a time context. "We usually sell out before 10:00am." "These are our most popular items and we only have a few available today."

Easy does it. Let her know you can take checks, cash or credit cards. Then, with as much confidence as you can muster, offer to, "Let me write a receipt for you while you make out your check."

Second...or third time's the charm. Don't feel as if you failed if you don't make a sale the first time a prospect walks away without making a purchase. Make it easy for her to come back. Ask, "May I put you on our email list for a (choose one) free farm newsletter, free monthly recipe, or advance notice of specials?" Following up like this subtly and psychologically establishes that your business is viable and you will be there next month or next year if it takes that long for her to decide to buy.

No. If someone tells you "no," find out why. Perhaps the timing is wrong, or there's something in your packaging or delivery that you could change or improve upon. Ask what it would take for them to reconsider. Once you've made changes, offer a sample to get the person's feedback.

Enthusiasm, focus, and persistence will eventually win over that initially reluctant prospect. This will encourage their repeat business and referrals of like-minded friends.

Anger management: how to handle criticism

No one likes to be criticized or told something they believe is wrong. When this happens there are two responses—you can avoid the conflict by drawing up into your shell like a turtle, or you can fight back. Either option is a lose-lose situation with your customers.

The only winning solution is to confront the person causing conflict and try to agree on a solution. Ready? Take a deep breath and relax.

Stay calm. When we get angry we get rigid and defensive. It is impossible to get anyone who is angry to change his mind because his heart, not his head, is in control. A customer at this stage is rapidly on the verge of becoming an ex-customer and a hostile one to boot. Whenever you get a phone call or face-to-face angry customer, the first thing you have to do is to get her thinking again. Take a deep breath to calm yourself. Ask her to detail exactly what she is upset about. Make her think! Then do whatever it takes to make her happy. Say something like, "Let's see if we can come to some kind of agreement." Or, "Let's see what we can do to fix this."

Disarm. Years ago I took a self defense course. One of the first things to do if you are attacked is to go slack—as if you have fainted. This makes your body dead weight to the attacker and much harder to handle. Maybe it originated as a Zen thing. When someone attacks you verbally, don't feed their anger. Let their unsavory words and attitude hit a blank wall. Smile. Take a deep breath so that you *can* manage a smile. Lower your voice in contrast to theirs. Feed their bitterness with sugar.

Butter them up. One way to make a disgruntled customer feel important is to bend the rules for them. You might say something such as, "Normally we don't do this, but I think we can work something out by....(fill in the blank)."

Getting to know your customer. Give the prospect

Inner Strength

If you can start the day without caffeine or pep pills,

If you can be cheerful, ignoring aches and pains,

If you can resist complaining and boring people with your troubles,

If you can eat the same food everyday and be grateful for it,

If you can understand when loved ones are too busy to give you time,

If you can overlook when people take things out on you when through no fault of yours, something goes wrong,

If you can take criticism and blame without resentment,

If you can face the world without lies or deceit,

If you can conquer tension without medical help,

If you can relax without liquor,

If you can sleep without the aid of drugs,

If you can do all these things....

Then you are probably the family dog.

Posting from the Internet

more information so that they won't see your product as a threat to their previous buying habits, and start to understand its benefits to them. Sufficient information will reduce any uncertainties they may unconsciously perceive when it comes to change. Address their fears. Put your product in the context of the unthreatening past. "Your grandparents ate like this, pure and simple." Explain how cooking methods are no different than what they are already familiar with.

Win-win. If you view business as a win-lose conflict, for the business to win it must lose the customer. If you let the customer win, you get to keep her and probably will get four or five new customers from positive word of mouth. To prosper you need a win-win scenario where both you and your customer win.

Information. Win-win scenarios start with very clear information that creates expectations in your customers that can be fulfilled. If you are getting a lot of angry customers the fault lies somewhere in the information you are giving out. Reconciliation of expectations and information is hard painful work and a continual task. Change that is forced or comes too fast seldom lasts. People need time to gather the information they need to alleviate the anxiety naturally incurred by change.

Make notes. Most likely you will go over the conversation in your mind long after the angry encounter. Rather than dwelling on the exchange endlessly, make notes. What did they say and how could you have given a better response? What caused the incident and how could it have been prevented? What did you learn from the experience? Writing it down will help you vent your pent up, unexpressed feelings (don't break any pens over it), and can guide you to a more confident response if and when it happens again. If you have a procedures manual for your farm, put this in it to help anyone who may be assisting you with sales when you can't be there. One day this will be history that bears remembering.

Let go. Some people will carry double chips on their shoulders as if they're gold-starred epaulettes. Being angry most of the time is second nature to them. Often this is a defensive tactic so ingrained they don't even realize how that comes across to others. It may be a sign of their insecurity—they attack so they won't *be* attacked. If your firm, yet gentle response has no affect, let that customer go. You will be happier for it because negative people can sour everyone around them. At the *Stockman Grass Farmer* we will return subscription money to griping customers rather than put up with their continual complaints. Always reject them politely, "Perhaps you'll be happier with another supplier."

Reward. No, not them. Yourself. After the encounter is over and the angry person leaves (hopefully happier and satisfied), give yourself a reward for handling a sticky situation. It will lower your blood pressure and keep you from dwelling on the unpleasant encounter. Reward yourself with some forbidden ice cream—a whole pint if necessary. Go shopping and buy something nice or silly to make you smile.

Educating your customers

People love to talk about food. Recently while exercising to get rid of those extra pounds, my water aerobics group was discussing food. Once every four to six weeks the group meets for breakfast in order to plan where to go for lunch.

If you're like me, sticking with generalities—recipe talk, what's new on the restaurant scene—is safe ground when you're talking with a group of people. When the "N" subject of "nutrition" enters into the conversation, I find most people are clueless about truly healthy eating. They're more comfortable discussing the merits of various diet schemes and will chant a litany of Diet-This, Low-Fat That, and Sugar-Free Whatever convenience foods they've found on the grocery shelves. Just because you're eating cottage cheese with one of those chemically produced artificial sweeteners doesn't mean you're giving your body what it needs.

Converting diet-conscious, convenience-food lovers into regular customers won't happen overnight. It's like the novelist who said, "Oh yes, I'm an overnight success. A *ten year* overnight success."

Standing on a soap box to deliver your message is hard for anyone to swallow in one setting. Yet, because you care, you want to help people eat healthier.

In his book *Cash Copy*, Jeffrey Lant, PhD, noted that it

takes seven touches over an 18-month period for people to notice your message. Most people won't register the first few "bites." They see your stall at a farmers' market. They drive by the farm and notice the name. They see someone carrying a shopping bag with your logo. They read a story about your farm in the newspaper. They see a sign you've sponsored at the Little League ballfield. They hear you doing an interview on the

How to be a savvy shopper

Not all of your educational messages need to be about grassfed products. Share grocery shopping tips with your customers. Turn this into take-home tips—even for those who don't buy (yet) from you.

1. Make a list to avoid impulse purchases.
2. Shop the edges of the store where the fresh goods are located.
3. Realize that cut up vegetables can cost as much as five times more than whole items you cut up yourself. The same thing applies to shucked corn. Convenience costs.
4. Take a calculator to figure the most economical sizes to buy. Buy by unit price.
5. Avoid products stacked at the end of the aisle, which can cost 30% more than the same items on the shelves, although sometimes they are the better deal. Compare before you buy.
6. Products at eye level cost more than their cheaper cousins on bottom shelves (and sometimes higher shelves).
7. Just because items are sold three to a pack or in a group of five doesn't mean you can't get just one for a third or fifth of the cost of the group.

radio. Best of all, a customer tells them about you.

In a similar manner, to educate consumers, you have to be persistent and use a variety of ways to present your message.

Love notes. Have you ever dropped a love note into your child's lunch box or hidden a message in your spouse's suitcase? Create a Did-You-Know? list of facts about the benefits of grassfed products. Duplicate the list and separate

8. Store brands are generally cheaper than well known brands.
9. Resist the urge to toss in those items displayed near the cash register. You will pay for the mark up.
10. Use coupons and store bonus cards. Don't overlook newspaper ads that say the store will match anyone's price.
11. The least crowded time to shop is between 9pm and 8am or on a Wednesday. Another slower time is during dinner time—5pm to 9pm.
12. Exotic fruits move slower than apples and oranges and so are not as fresh.
13. Have you ever found the baby food or batteries in locked cases? Thieves resell them on the black market along with quit-smoking products, smokeless tobacco, razor blades and cough and cold medicines.
14. Grocers market to children by putting the candy and flavored cereals at kid-eye-level.
15. It will take longer to process a coupon that doesn't have a scan, such as those newspaper ads that say they'll match prices.

This kind of education aids in your customer's trust because it's free. (Feel free to copy this list or create your own.)

each item into a message you can slip into the bag with each customer's order. Perhaps add a recipe on the other side so that the consumer will keep the note or post it on the refrigerator. Change the message each week.

Wendy Taggart of Burgundy Pasture Beef in Grandview, Texas, not only has flyers with recipes, but tucks other educational flyers into each sale package. One flyer (an 8.5x11 sheet of paper divided into 2 pages each side—when folded has 4 pages) explains the benefits of Vacuum Seal packaging, the recommended type of cooking utensils, basic culinary techniques, how to braise and how to brown a roast, how to cook premium steaks by pan frying and grilling, plus some recipes. Every bit of the information is worth keeping for future reference.

Consult cookbook classics and create your own info-mercials.

Signage. Use signage in the farm store or farmers' market to explain the benefits. Alternate signs periodically so that customers will look for your messages and learn something new each time. Hang a bulletin board and change out articles about healthy lifestyles. Let this be the place your customers go to learn what's happening in the grassfed world. Post it on an easel in front of your booth so that passers by can read it without feel obligated to buy something (at least not the first time).

Library. When a customer expresses an openness to learn more, give them reading material to help them become informed consumers. Jo Robison's *Pastuare Perfect* or Michael Pollan's *In Defense of Food* and *Food Rules* are books that should be on everyone's shelf. If you don't want to stock enough of these books to sell, consider buying multiple copies to create a lending library to educate your customers without them incurring an extra expense. The *Stockman Grass Farmer's* "Grassfed" brochure contains just enough facts without overwhelming your prospects, and it's priced so that you can use it as a give-away. At the very least create a Recommended

Reading List with brief descriptions of books or websites that can flesh out your message.

Speak out. The program chairperson of civic and social clubs is always looking for new people to present a program. Let your customers know that you are available for speaking engagements. When you offer to make a speech for free, they'll be eager to book you. Use that opportunity to talk about the benefits of grassfed products and back up your message with a Crockpot sampler that you offer the audience.

Cooking class. Give a cooking class. While you're demonstrating the multiple uses of grassfed products and the ease of cooking toss in educational facts. Send everyone home with a list of benefits. Be sure your farm's name is on everything you hand out. If regulations permit, do a cooking demonstration at the farmers' market. Use your church's kitchen. Contact the Convention & Visitors Bureau to market your cooking demonstrations as a program for wives whose spouses are attending a convention. Capitalize on childrens' impressionable years by presenting a school program for the Home Economics class. Work with a caterer or a gourmet cooking shop.

Going green. Link your wagon to groups concerned with environmental sustainability. Let them know how grassfed products fit their cause by saving fuel otherwise wasted on transportation costs; by keeping money in the community; by saving water with dense, living pastures; by harvesting sunlight through grass; and by saving family farms from developers who clear cut and cover the land with concrete.

One small step. Remember, Rome was not built in a day. Neither should you expect grand results immediately. It's hard to break old habits, even when you know doing so is better for you. Some people will be like a friend of ours who would stick her fingers in her ears whenever we tried to tell her the truth about confinement chicken houses. Nothing we told her would sway her from her favorite meat. At the time grassfed chickens were not available locally.

The grassfed challenge. Making changes, particularly when it comes to healthier eating choices, can be overwhelming. Challenge consumers to make one healthy change a month. Offer enough suggestions to let them pick and choose: Make their own mayonnaise or salad dressings (offer 15 Minute Recipes). Read labels and eliminate foods containing soy or excess salt. Cook with fresh or frozen vegetables rather than chemically-added canned vegetables. Always choose grassfed over grain fed-dairy and meat products.

Building community

To paraphrase John F. Kennedy, ask not what your community can do for you, but ask what you can do for your community.

Building community means being involved in the place where you live. It's proactive. It takes the pressure off the all-too-natural fear that someone will *ask* you to do something for them.

Building community helps others, and the ripple effect helps you and your farm business by being a working example of good business practices, by providing quality products for the health of those who live near you, by creating a sustainable system that is good for the environment, by being a good neighbor.

Building community can help you build name recognition among your customers and prospects. Buying a sign for the fence of the Little League Baseball park is a start, but it's not nearly enough. Volunteering your time, energy, and sometimes your products for things you believe in can build networks within your community.

Have a cause. When it comes to selecting grassfed over grain-fed products, women are more likely to make their selection because of their concerns about the environment. My friend, Beth Greer, author of *Super Natural Home* believes in providing the best food for her family. She started campaigning for organic foods and grassfed hamburgers to be served in her daughter's day care. As a result, healthier meals moved "up the food chain" as her daughter entered primary school. Other schools in her area adopted that model.

Following Alice Waters' example, offer to set up an Edible Schoolyard in your community to teach children the value of fresh, locally grown vegetables. Working with your community, you could do the same with a flock of baby chicks raised by school children on your farm. Children love gathering eggs.

Sponsorship. Is there a charity or cause you believe in to offer sponsorship by supplying your farm products? If not, create your own reason to Celebrate the Grassfed Farm by donating fresh and canned goods for the local food pantry.

Women are more likely to buy holiday items from retailers who support social issues. Studies have shown that 65% of women shoppers will purchase something if they know even a small percent of the sale goes to a cause they believe in. By contrast, only 54% of men are motivated by sales for a cause.

If two products are exactly the same, except for the packaging, women shoppers will select the one that supports a

What builds a sense of community?

Bo Burlingham identifies four steps for building community.

1. Integrity. Your customer knows your business is what it appears and claims to be.

2. Professionalism. Your business does what it says it will do.

3. Direct, human connection. How you do business creates an emotional bond based on mutual caring.

Small Giants

cause. "Women respect companies whose support is consistent, substantial, and ideally interactive with what matters in their lives," wrote Mary Lou Quinlan in *Just Ask a Woman.*

Does your budget allow you to buy uniforms—with your logo—for a Little League or soccer team? (Think twice before you invest. Are your customers parents of the players?) Can you set up a booth to sell sloppy joes at the ballpark? How about offering to supply a meal for Habitat for Humanity workers? Pack them in brown bags and stick your label on the bags. Or sponsor a Brown Bag luncheon at the library or outdoor concert.

Speakers circuit. Joel Salatin helped build a database of customers for Polyface farm by speaking at Rotary Clubs. He always brought a Crockpot of samples for the audience to taste. Offer to speak to local civic and business organizations. They are always looking for fresh program ideas.

Worth noting. Let consumers know about your "corporate citizenship" good deeds through your flyers, displays, websites and when selling. I was cleaning up our two-person pedal boat for our homeowners' July Fourth fireworks show when I remembered that Dawn dishwashing liquid has a phrase printed about helping clean up wild ducks. Without looking further that implied it was biodegradable and wouldn't harm our lake environment.

Paul Newman's "Newman's Own" line of salad dressing was created as a vehicle for his charitable giving. You can read that on every package and ad. M&Ms candies had a package of pink and white M&Ms to tell customers of their donations to help fight breast cancer.

Packaging. Doesn't it irritate you to buy a 4-inch package that contains a 2-inch item? You end up throwing away a lot of unnecessary cardboard, often layered in shrinkwrap. Even worse are those plastic display cases that require industrial scissors to break out the much smaller item inside. How many trees were lost just to make sure that two-inch item had twice as much shelf space as it needed? How long will that mangled plastic remain in a landfill?

Think "environmentally friendly" for your packaging. Give your best customers a reusable cloth shopping bag with their mega bucks purchases.

Recycle. At the *Stockman Grass Farmer* we print on the blank side of junk paper. Those proof reading for an issue often become distracted by reading something interesting that had been printed on the other side.

Remember, nothing you do will come to fruition overnight. Similar to the Edible Schoolyard, you will be planting seeds of living education for the future of your community.

Where to sell

Farmers' markets

Farmers' markets are growing in number and popularity coast to coast. Since 1994 they've increased 13%. In the urban environment they've become the new scene for social networking. They offer a welcome change and often better options from what's available in our generic grocery stores.

Most direct marketers will start selling through farmers' markets. These venues pre-select customers who are seeking fresh, local products. They can serve as a testing ground before opening an on-farm store. As teaching facilities they can give your children or employees hands-on experience. They are a service to your community.

And yes, they are a pain in the neck.

Expect to spend at least an hour loading things up, another hour setting things out, another hour packing things for transport home, plus the hours you'll spend driving and selling, often on your feet. It all adds up to a long and exhausting day.

Some marketers limit these headaches by only going once a month, selling for a few hours, or making the effort worthwhile by delivering pre-orders at the farmers' market site. If you're not going to be there every time, make sure you post your Hours of Operation for your customers' convenience. Judi Sedor of Newgate Farms in East Granby, Connecticut, divides and conquers. She sells her produce, cakes and jams at nine farmers' markets by sending her five children to different ones.

Even though they're a lot of work and your customer base has moved beyond the need for you to sell at farmers' markets, from a community-building stance, it's good for the market and your farm to keep a presence there. Consider using your farmers' market position to promote an on-farm event—a field day, open house, festival or other venue to attract customers to your home base.

If you have more than one option, look for markets with the best locations. That translates into busy foot traffic and sales. Markets centered around an elderly population may not be as lucrative simply because many seniors don't eat much meat. How often have you seen these folks in a restaurant splitting a meal?

Benefits. Until you're ready to move things closer to home with a farm stand or store, remind yourself of the benefits of farmers' markets. They provide a means of building a loyal, regular clientele. Capture as many customer names, addresses and contacts as possible to establish a personal database.

Fresh sales. David Evans of Marin Sun Farms in the San Francisco Bay area cites another advantage. Fresh meat

Start your own

American Profile, a newspaper magazine insert, profiled 14-year-old Lindsey Brenkus who was so taken by a farmers' market in Mount Vernon, Ohio, that she became the driving force to establish a farmers' market in her hometown of Bellville, Ohio. Not only does Lindsey mark off vendor sites, stay up on and enforce the rules (no crafts), but she even hired her younger brother to help with the advertising. For $1 an hour she had him dance on the sidewalk with a farmers' market placard. He then decided he needed a raise.

sales at the San Francisco Ferry Terminal Farmers' Market were double the price of frozen product. What doesn't sell can then be frozen or used in an on- or off-farm restaurant or store.

Variety. Farmers' markets offer the best variety (heirloom tomatoes, eggs from truly free-ranging hens, wild huckleberries) and freshest products for consumers.

Food safety. Licensed vendors with permits from the Health Department equates to high quality, safe, clean and healthy products. As one farmer quoted on Seattle Farmers' Market website stated, "Farmers who sell direct have a great interest in making sure their products are safe and wholesome. If that was not the case, they would stand to lose their customers and/or their farms."

Stretching food dollars. A study conducted by economics students at Seattle University and reported in the Neighborhood Farmers' Market Alliance newsletter showed that on average, prices at the farmers' markets were 86% of QFC (a grocery store) prices. Over two weeks, the students studied the University District Farmers' Market, Whole Foods and QFC. Average prices per pound were $2.37 at the farmers' market, $2.59 at Whole Foods and $2.97 at QFC.

Interestingly $20 will buy 8.4 pounds of produce at the farmers' market, 7.7 pounds at Whole Foods and 6.7 pounds at QFC—proving food dollars go farther at the farmers' market.

Another study by Ohio State University and published in the *American Journal of Agricultural Economics* showed that farmers' markets' shoppers were willing to pay almost twice as much extra as retail grocery shoppers for the same locally produced foods.

Get one started. If there's not a farmers' market within reasonable driving distance from your farm, you can spearhead one for your community. There are plenty of online websites to tell you how to get a farmers' market started. Simply Google "farmers markets."

Practical issues of farmers' markets

Whenever we drive to Jackson, our state capital, we always see trucks loaded with sweet potatoes parked on side roads and the farmer sitting in the shade nearby.

Selling at farmers' markets seems as if it could be almost that easy—set out your stuff and start making sales.

However, there are rules and regulations that vary from market to market (some prohibit crafts, pets, and—gasp—some don't allow meat sales) as well as by city, county and state. Google "farmers markets" and you'll find more information than you need and can pinpoint your area, so we won't attempt to cover all of the variables. Contact the Department of Agriculture for specifics in your state.

There are some practical issues you need to consider.

Taxes. First understand and follow the tax regulations for your area. In some places food isn't taxed. In others it is. If you live where food is taxed, add it in to the prices you post, and round it up so that your customers and you don't have to lose time finding and making change.

Electricity. Some markets have power outlets for those of you using a freezer. Otherwise, you'll need to invest in coolers and cold-packs. If you do have access to electricity, you could decorate your booth with strands of chili pepper lights or colored blinking lights, anything to draw attention to your display.

Signs. Order a banner sign to clearly identify your booth. This helps customers easily and quickly find you if you have to shift locations within the market.

Your signage should be crisp and clear. Use a sandwich board in front of your booth to call attention to a particular feature. Change this with each visit to pique your customers' curiosity.

Marlene Groves of Buffalo Groves uses a large banner and a decal on their freezer as identification. She has blown up and laminated price lists, plus offering customers the use of smaller, menu-like laminated price lists. Chalk or white-boards can be useful to erase products as they're sold out. Even better would be to strike through and write SOLD OUT so that potential customers can see what they missed.

Divide and conquer. David Evans of Marin Sun Farms rented a large booth at San Francisco's Ferry Terminal. He admits the 13-hour Saturday market is exhausting, but profitable. He typically brings 15 coolers of meat and sells all but two or three. It takes two employees a full day just to price items the day before. Three people staff the booth at all times, rotating with other employees to avoid burnout from being upbeat and "on-stage" for customers.

Share your shade. Some marketers don't use pop up tents or picnic table umbrellas. If you do, set aside an area or plan to have an overhang that shades your customers. Use weights to keep your tent from blowing away. Or you might be able to use bungi cords to tie it to your vehicle.

Sampling sells. If your market allows, giving away samples sells, particularly if you have the added lure of the aroma of grilling meat. Be particular when it comes to hygiene. Use plastic gloves. Keep the samples covered to avoid pesky insects. Skewer meat or cheese with toothpicks to avoid your customers' hands touching the other samples. Hand the skewered item to the child or adult to keep the other samples clean. If samples are small, place each in paper cups, the kind used in restaurants for ketchup and mustard.

Rewards of "going green"

Here are some contributions customers make by buying at farmers' markets:

Face time. Farmers' markets allow customers to meet their farmers. By visiting the farm they can see the healthy conditions their animals are raised in. They can see the pride in the person who shepherded the animals from which their meat and milk products come. There's no commercial corporation or middle man to hide the producer.

Save the farm. By supporting local farmers, they are keeping money in the community and keeping the producer viable on the farm. This allows the small family farmer a better income and standard of living than from commodity sales.

Save the land. Supporting the farmer and his farm helps keep pastoral landscapes alive. Healthy farms help block the development of commercial building projects, which often start by cutting down all the trees and removing precious topsoil to level the site. Trees and grass create cooler environments than concrete buildings or paving. Plus, healthy soils hold more water.

Healthy environments. Supporting the farmer saves fuel that would otherwise be spent in long-distance trucking fees—lettuce from California, oranges from Florida, beef from Kansas. The average steer is trucked 1500 miles in its lifetime before reaching the grocery meat counter.

Good practices show customers that you care about food safety, which as Janet Majure reported in *Growing for Market* newsletter, this guards against greater government regulations, which vary from state to state, city to market.

Bag your supplies. Like frequent travelers, keep a bag of supplies packed between market dates. Useful items will be a broom, light rain jacket, clips, clamps, scissors, tape, rope or bungi cords. Signage and printed materials can be stored in an artist's portfolio—these are over-sized, heavy-duty, cardboard briefcases. This will keep them together and avoid damage during transportation. Don't forget a trash can.

Practice makes perfect. Develop a routine so that everyone knows what their job is when it's time to set things up and take them down. Get there early so that you're fully set up when the market opens and not hassling with your materials while trying to sell something to a customer.

Transactions. Since most people don't carry a lot of cash, and you want to make things easy for your customers, consider accepting credit cards (realize you'll lose a percentage of the sale in processing fees) or accepting checks. Accepting credit cards (the more you can handle the better) encourages impulse buying and larger orders. And it shows customers you care by accepting cards. Ask for an ID and get a phone number. Marlene offered this tip: "Sometimes the manual credit card machines don't read the numbers so great and if you key in a 1 and it's really a 7 the charge does not go through."

Most people are honest, and you won't lose money accepting a check. You'll know if someone writes a bad check since market vendors will quickly spread the word.

Keep small denominations on hand to make change. Or, offer something additional as compensation. Our egg vendor ran out of change, and gave us a selection of cut flowers instead. And once when the last tub of yoghurt had already been sold, the dairy farmer let us have the one he had been using for samplers. No charge. We'll be back.

Making your booth stand out at the farmers' market

If you're going to take the trouble to haul your products to farmers' markets, make your display area a magnet for customers. Even if you are the only one selling grassfed meat or dairy products, having an enticing booth will demonstrate professionalism. Think of it as if you're decorating a room.

Colors. The Wagner Institute for Color Research in Santa Barbara, California, found that blue discourages spending money. Cool green encourages spending, however a warm green carries an institutional quality. Yellow can attract people to your booth, but the color makes people uneasy. Orange implies something cheap, like fast food. Black and red are bold, power colors but should be used in moderation for impact or accent. Couple those power colors with natural tones, gray, white or tan.

The website for Glidden paint says that colors can reflect emotions, love, determination, as well as conflicts, challenges and frustration.

Here are some hidden messages in colors:

Black, gray, and brown = all suggest negativity.

Blue-green = persistence, confidence, and even obstinacy.

Dark blue = depth of feeling, tranquility, love and affection, but does not work well for food.

Gold = expensive or high quality.

Gray = masculinity.

Green = natural.
Maroon = pricey.
Orange-red = forcefulness, desire, excitability, domination.
Pink = femininity.
Red = this attention getter works well with low prices.
Violet = taste and grace.
Yellow = spontaneity, originality, exhilaration.

Be different. If everyone else is using pop up tents for shelter, use patio umbrellas at your area. Two or more can suggest a tropical or beach theme, and you can use them in your own backyard at home.

Food displays. Since most of you will be selling frozen products, you may want to photograph various cuts and post these near the price chart.

Marlene Grove uses "make believe meat." For their Bucket of Buffalo offering, she wads up red paper, puts it in a cellophane bag and sticks their label on it. For their Steak Plate gift pack, they put a *Buffalo Nation* book, a package of painted styrofoam steaks resembling New York style cuts on a rustic, miner's tin plate.

At the Birmingham, Alabama, Farmers' Market, the Jones Valley Urban Farm uses feed sacks as tablecloths beneath baskets filled with their cut flowers and vegetables. It adds an old-fashioned tone to their booth.

Decorate with drying flowers or hanging herbs—branches of rosemary, braids of garlic, strings of red chili peppers. Aromatic herbs draw people in.

What do you see?

Women see a greasy kitchen counter and think, "That needs cleaning."
Men see a place to set the dead car battery.
Marni Jameson, a Denver, Colorado, columnist who writes about home and family issues.

Less is more. Depending on the item—individually wrapped cheese wedges, for example—let them spill out of an overturned basket. This gives the impression of bounty when in fact supplies may be limited. This is particularly appealing if you sell vegetables in your product mix.

Use racks, baskets, boxes or crates to add color, height, variety, and interest to your table display. Just don't stack things so high you can't be seen.

Signs of the times. If you have a logo or distinctive type style, keep it consistent with your letterhead, newsletter, shopping bags, order forms and signs.

Use key words to quickly attract attention: pasture-raised, grassfed, drug free. Red or black on white is easiest to read. Turn your pickup or panel truck into a rolling billboard.

Public relations. Marlene Groves has laminated nutritional information posters from the National Bison Association and keeps press clippings available for customers to read. "Believe me, they do read it," she explained. "You see sometimes they are just looking for information, then a few weeks later they buy."

Create a lure. Do something to get people to stop by your booth even if they don't buy anything. In the early morning, be the one who has free coffee and cake available. Make a different cake each time. Have copies of the recipe ready to give anyone who asks about it. You can do the same in the farm store.

Looking good! Keep your display area clean, neat, and clever. Neatness counts. Be creative.

By making your stall a stand out, your booth may be the one selected when the media needs an attractive photo to illustrate their TV segment or newspaper story.

Meat displays

Peter Burns from Ridgway, Pennsylvania, wrote that farmers' markets—his family sells at three—form the foundation of their marketing plan. They sell eggs, beef, pork and chicken from small portable coolers. They hang colored signs with pictures to represent the contents of each cooler and have even used stuffed animals as attention getters. "These things do work," he admitted, "but we would like to do better."

Transparency. Trusted clients won't mind butcher-wrapped meat. They've bought from you before and they know the quality they're getting sight-unseen. However, new customers like to see what they're buying. Just as in a supermarket meat case they want to select for marbling and general appearance. Invest in clear packaging. One exception is that ground meat virtually looks the same and can be packaged in opaque wrapping.

To minimize the time the cooler is open and packages are moved around (even vacuum packed product can be damaged if shuffled too much), describe the contents. For example, you might write "heavily marbled," "two-inch thick," "boneless." Price lists outside the cooler also help minimize the number of times the cooler is opened, which can lower temperatures particularly on hot days. Rather than letting the customer re-arrange your cooler's content, pull out a couple of items to give them a choice of one over the other.

How to photograph food

Lighting. If you don't have access to a professional set of lights, look for even lighting outdoors. Cloudy bright days are better than full sun, which can create deep shadows and bright hot spots. Photographing in shade is one solution, although be careful of the source of the shade. Photographing beneath a tree risks giving the food a green tint. Watch out for color tinting from canvas umbrellas.

A problem with photographing outdoors is that the food can attract flies. Have an assistant ready to fan intruders away.

Indoors, make camera adjustments to compensate for greenish fluorescent lights or orange cast by incandescent bulbs. To work around these unnatural light sources use available light by setting up near a window.

Framing and focus. Less is more, unless you're trying to set a mood with a wide angle view of an outdoor picnic or indoor table laden with food. Get close to focus on the "star" product. Let everything in the background go soft.

Angles. Experiment with camera angles. Professionals will get down on the same level as the food to shoot head on.

Background and mood. Pay attention to the background. "Stage" the setting by putting appropriate props around the main item. If you want to suggest a hearty, ranch-style meal, use a barnwood background or table. An elegant dinner will have crystal glasses and linen napkins. A checked tablecloth suggests picnic fun.

Photos of food. Ethic restaurants often use photos to aid diners in placing their orders, and of course you'll find them at fast food restaurants. We encountered photo menus in Peru where guinea pig is a specialty item. The first night we identified the Spanish name on the menu and got up the nerve to order guinea pig we were told a two hour notice was required (no doubt enough time to catch, kill, skin and cook one). The second time we decided to try it was in a restaurant with a photo of a plate containing the cooked guinea pig, head on and legs up. Needless to say, we saved the lives of at least two guinea pigs on that trip.

Hire a professional photographer to take photos of your mouth-watering meat— grilled steak, slices of roast, skewered kabobs—with complementary side items set on an attractive plate. Don't use a blue background, which tends to make food look unappetizing.

Photos of the farm. Blow up pictures of your happy animals grazing lush, green pastures. Customers care about the humane treatment of animals. They want to know their meat came from pretty, neat farms in a beautiful environment. Include photos of your family interacting with the animals. This kind of photo makes a strong statement and can often be all that's needed to persuade potential shoppers to buy your products.

Signage. Promote value by stating the benefits of grassfed products. No hormones, no antibiotics, no crowded conditions. Use a professional sign company to give posters a clean, crisp appearance. Create flyers that can be copied inexpensively and given away. Have plenty of business cards on hand. Remember, four-color brochures and advertising can't sell a product that's only "good enough." Strive for the best quality possible. Quality beats quantity every time.

People care about the welfare and humane treatment of animals. Farmers who are part of the Animal Welfare Approval program can request a range of banners, signs, brochures and labels from the organization AnimalWelfareApproved.org. This group can now certify farms in Canada.

Don't forget to create signage for your vehicle to give people stuck in traffic something to read and remember.

Aroma. While a picture may be worth a thousand words, the aroma of savory grassfed products cooking speaks for itself. If at all possible, prepare samples onsite to entice customers to taste the wholesome goodness you have to offer.

Professionalism

Recently visiting a farmers' market (the location will not be named to avoid embarrassment) created a heated discussion—not about what was on offer, instead about the lack of professionalism of the vendors and the market manager. This can work to your advantage if your products and professionalism eclipses everyone else's. However, just as a customer's bad experience with grassfed products that are only "ok" and not "exceptional" can hurt the producer and everyone in the industry, the lack of professionalism reflects badly on everyone at the market.

Box-farmers. Many markets with lax restrictions allow "box farmers." These are people who buy goods wholesale for resale. In most cases they're not farmers. They may be allowed in for the purpose of building up the size of the market, and on a positive note, their presence brings in more clientele. However, including such vendors can erode the consumer's trust in the market and its goods.

"Yes, it's frustrating to (be) undercut by a low-priced alternative offered by (someone) who can't even be bothered to peel the bar-codes off his wholesale cases," stated Paul Schwennen. "It needs to be clearly stated who is the farmer and who isn't. Our customers come to the market to buy from us, not to get bananas and out-of-season strawberries."

Butch Tindale of Homestead Heritage Farms in Elm

Mott, Texas, put it this way, “Most people don’t want to get their hands dirty, but want to shake hands with someone who does.”

Counter action: If the market manager doesn’t enforce identity traceability of its vendors, make sure you do. Use every tactic you can to make it clear you and your family are the sole producers of your products. If you carry someone else’s grassfed products, make it clear who the products came from and stress their like-minded, sustainable practices. If a neighbor’s products don’t meet your standards, don’t mix their products with yours.

Daddy made me do it. Market managers can’t control who’s in the booth. You can. Realize that farmers’ market shoppers are not run-of-the-mill, bulk-store shoppers. They care about their food. They want fresh, local, healthy products. They are informed and will ask a lot of questions. If you are offering something unique (and you should be if you’ve created and filled a niche no one else has), take care with the person you send to man the booth.

At this nameless market, we zeroed in on two of the most unique (to that farmers’ market) products—goat cheese and cow-dairy products. At both booths we wanted to know more: Were the animals fed any grain? How was the milk processed? In both places the sellers were clueless. They had just been sent to work the booth, and it was obvious they didn’t enjoy being there. With the cow dairy we were able to pick up a brochure that answered some of our questions. Even though we liked the products, the people taking our money did not inspire us to support them by regular visits.

No one enjoys shopping with disgruntled business people. Given a choice between two identical products, customers will always buy from the most likeable person. Personality matters. A simple smile goes a long way.

Counter action: There will be times when the main marketer in the family is unavailable. Choose your substitute carefully. Call on a neighbor, friend, or family member who has a pleasant, enthusiastic personality. In general, women are better marketers than men. Coach them on the most important benefits of your products. Create a Q&A "cheat sheet" of frequently asked questions so that they will be prepared on how to respond. Load them down with brochures that explain things they aren't well versed in so that they can hand these out. If they can't answer a customer's question, be sure they say, "I'll find out and have that answer for you next week." And here's a trick that can save the day—once—have the substitute admit, "This is my first day. Please bear with me. I'm still learning."

Signs of the times: The lackadaisical market—now in

Farmers' Market professionalism

With 100 vendors and six weekly markets at various locations, the Farmers' Markets at Fayetteville, Arkansas, is organized for professionalism.

Each vendor has an 8-foot space, too small for pop up tents so they use colorful patio umbrellas. To encourage participation in the weekday markets, vendors receive points. One for each year of membership, one for participation in the Saturday market, two for being at a weekday market, and one point for each $10 gross sale. This encourages an active market with a waiting list of vendors wanting market space.

Fees are $5 annually, plus 5% of weekday sales and 7.5% of Saturday sales. All sellers must be farmers who produce their own goods, no box-farmers allowed. The market manager visits applicants' farms for verification.

The pay off is 4500 shoppers on a peak summer Saturday.

its third season—still has little to no signage. In an article by Isabelle Groe posted on Seattle's farmers' market site, only 2% of farmers in British Columbia participate in the farmers' market. Some don't trust that they'll be successful there or that the market will last. Many markets only have temporary locations and a move will cost everyone customers. Others just don't want to invest the time and energy it takes to sell there and think any time away from the farm is detrimental to their production. Set up costs keep others away, and many just don't have the people skills or confidence to sell their goods face to face.

Counter action: Just as a successful business takes ten years to really crank out profits, selling at farmers' markets takes an investment in time. The best way to develop confidence is to grow slowly with the number of customers who increasingly notice your booth and seek you out. Don't give up at the first sign of dwindling sales. Ask yourself why that's happening and what you can do about it. Remember, capturing customers' names for your database is priceless.

Appearances: Some vendors back up trucks and sell directly from the truck bed. While this identifies them as "truck farmers," it also suggests they're too lazy to set up signs, or assemble their products in an attractive manner. Like children watching the clock for the final school bell to ring, these vendors are poised for flight as soon as their bins are empty or foot traffic dwindles (whichever comes first).

Counter action: Wearing farmers' overalls can be appropriate as long as they're neat and clean. If more than one person is working your booth, wearing some kind of uniform will add a touch of professionalism. Name badges indicate you want people to get to know you, after all, you want to be *their* farmer.

Lack of service: Yesterday I took my antique sports car to the dealership for its annual oil change and check up. Although I could see two service reps in their windowed cubicle, I stood outside my car for several minutes before anyone came out and asked me what I needed. A quicker response would have made me feel important, as if they cared about the customer. Then I held my breath while they gave their prepared "up-selling" speech telling me I needed more than an oil change.

Counter action: If someone who's not your regular customer expresses the slightest interest in your booth, be ready to reel them in as quickly as possible. However, choose your words carefully. Say something like, "Let me know if I can answer any questions for you." That keeps decision-making control in their court.

Making farmers' markets fun

Farmers' markets that are done really well can add vibrancy to a neglected downtown. People spend more time talking together at farmers' markets than they do in grocery stores. If you're going to go to the trouble of marketing there, make it fun.

Music, music, music. To draw a crowd, many markets schedule live entertainment during good weather months. Even in its first year, our little farmers' market let everyone know there would be a live band one night each month.

If your farmers' market isn't big enough to have an entertainment budget, create one in your own booth. Surely someone in your family—don't rule out grandparents—loves to play the banjo or guitar. Who can resist a child who sings?

Live entertainment. Maybe your children are too shy to perform face-to-face with their audience. Suggest they write and then perform a puppet show. Post a schedule of performance times.

The farmers' market at San Luis Obispo, California, has live jugglers (with fruit, perhaps?), bands and puppet shows. It's located downtown in a walkable area. Plenty of bustling activity draws a crowd, making it a social event.

Cooking events. Many markets feature regularly scheduled cooking events with chefs from local restaurants using the market's fresh products. Any excuse is cause for celebration—Berries in Season, Greens Galore, From the Pumpkin Patch—with a big blowout for the market's anniversary. Tie in to the market's events with suggestions on how to complement these cooking demos with your grassfed products.

Portland, Oregon's, farmers' market hosts an annual Grazefest featuring grassfed samples, tips, recipes, and demonstrations to kick off the barbecue grilling season.

Fun and games. Many markets purposely target children, knowing they'll have adults in tow. Some set aside a safe, supervised play area for youngsters. Others offer games and competitions.

The most creative farmers' markets have learned how to turn a problem into promotion. In late summer, everyone knows zucchini runs rampant. Joel Salatin said in his area of Virginia, people have learned to lock their cars rather than return to find sacks of zucchini someone has left on the front seat. The Seattle Farmers' Market is one of several that celebrates these out-of-control green squash with The Zucchini 500! Kids of all ages design and carve out vegetable vehicles and then race them. Best of all, the event is free, but the return to the vendors is priceless.

More free stuff for kids. It doesn't hurt to kid- and dog-proof your booth. Some vendors use pvc pipe as extenders on table legs to raise the height, but this is detrimental for physically challenged shoppers. That said, have something for children as giveaways—stickers of your farm's logo (with phone number), postcards of your farm. Kids love little things and the parents appreciate it too.

The Seattle Farmers' Market wisely targets children with regular events. In spring, they give free sweet pea plant starts...to children only. Several chef events are designed for kids to cook. And the anniversary celebration includes free birthday cake. One of their Chefs in Residence shows mothers how to make easy, tasty baby food using the market's seasonal, fresh products.

Lisa Agee from New Milford, Connecticut, sells 25

different soaps using goats' milk. To draw kids to her booth she has a small fenced area with a couple of her Nubian baby goats. While the children are petting the goat kids, adults are perusing her goat milk products, which include lip balm and bug spray.

At the Brookhaven, Mississippi, farmers' market one vendor has a large bucket of potting soil, seedlings, and small styrofoam cups. These are in front of the booth to attract little ones. They love taking home their very own plant, and planting their seedling keeps them safe and busy while Mom shops. (Be sure to keep wet wipes handy.)

Read all about it. Some markets feature a book booth —cookbooks generally, and books on healthy lifestyles, such as

What is local?

Beware of the word "local" in food advertising today.

Researchers at Ohio State University found that shoppers in that state consider anything that comes from Ohio as local.

Supermarket chain Giant Eagle defines local as being from the state the store is located in.

Kroger gives itself a two-state range for local.

Supervalu's Sunflower Market says local is a distance of 200 miles. That's the same distance Telluride, Colorado's, Farmers' Market allows for their vendors. Boulder's Farmers' Market must come from Boulder County, Colorado.

For Whole Foods Markets, it means the food traveled 700 miles or less, about half the normal distance of American food stuffs.

Stockman Grass Farmer, October 2007

Michael Pollan's *Food Rules* and *An Eater's Manifesto*. If your market doesn't sell books, add appropriate titles to your grassfed offerings at your own booth.

Don't go overboard until you've test marketed the reaction and sales of select titles first. Then contact the publisher (or author such as Joel Salatin—*Everything I Want To Do Is Illegal* and *Holy Cows & Hog Heaven* are directed at consumers) for their wholesale discount policy. You may be surprised to find out how few books it takes to get a wholesale price, then you can resell at the full retail price and pocket the profits.

Just as you should know everything about your products to answer any questions that arise, select books you love and have read, knowing that they fit well with your product lineup.

If you're selling in a major metro area, you may be able to create a book signing event with the author. Even in small towns, you could sell local church cookbooks and invite one or more of the prominent contributors to sign books.

Lagniappe. In Louisiana, the custom of giving away something extra or unexpected for free is a common occurrence called lagniappe. For example, order a dozen donuts, and you'll get 13. Your market, or your personal booth, can offer free cups of coffee or fresh squeezed orange juice during the early morning hours when foot traffic is heaviest. Plan your booth as a place customers head for first and give them reasons to linger awhile.

Too much of a good thing. Sometimes farmers' markets can become so famous they lose their original customers and focus, making it difficult for farmers to make a living from them. Due partially to its waterfront location, the Ferry Plaza Farmers' Market in San Francisco attracts more tourists, who purchase less, than residents. Tourists don't make weekly grocery purchases.

Its popularity has resulted in many vendors selling the same thing, with little distinction among the offerings. Fuel prices and distances from farm-to-market have caused other farmers to drop out and seek alternate selling venues for their products. Others have notified their regular customers that they will shorten their selling season there.

In New Orleans, the produce in the historic French Market—once chockablock with vendors of local pecans, Creole tomatoes, sweet potatoes, okra, spices, merliton squash, Creole cream cheese, strings of red peppers, Vidalia onions—has sadly dwindled down to a few food vendors surrounded by kitschy flea market and clothing sellers.

Incorporate farmers' markets into your overall marketing strategy. You don't have to be married to it forever, however, it can be the first lucrative step for direct marketers of grassfed products.

Load the wagon

Business experts say it is much easier, and less costly, to sell *something more* to an existing customer than it is to *create* a new customer.

We call adding things to your core product "loading the wagon."

Once a woman has "invested" her time and money in your product, after her first purchase she is more likely to stick with you due to brand loyalty. Cross-selling gives your customers more options to buy related products from you. Up-selling moves regular customers to purchase pricier items or more products.

Passion first. Rule number one is to be passionate about what you do. If you totally believe in your grassfed products you'll want everyone to know it and experience the taste and tenderness first hand. Selling will become part of your persona.

Your enthusiasm can rub off on your customers while it stirs your creative juices for ideas to make what you sell more imaginative. You're not just selling a lump of frozen meat. You're selling a main dish for a Sunday church picnic on the grounds or a roast to impress relatives at a family reunion or maybe just winning burgers for the neighborhood block party.

Stay true. The second rule is to stay true to your main product. My friend, Patricia, has created Kids in My Kitchen.

She teaches toddlers to teens how to cook. She also stocks food items that appeal to kids—novelty kits for making "Spider Web Pizza" and "Monster Punch" that comes with gooey Gummy Bear frogs. Non-food items are cloth shopping bags, munchkin-sized aprons and poofy chefs' hats—all imprinted with Kids in My Kitchen logos. By selecting products that complement her cooking classes, she has created a lucrative business. The additional items keep her income steady by bridging those off-week periods when no classes are scheduled.

"Nothing is original," said Seth Godin In *The Big Moo.* "Most composers, great or not so great, are working with the same musical alphabet of pitches. It's putting the pitches together differently that creates something remarkable."

Build a pyramid. Ancient Egyptians, Mayans, Chinese and other cultures built pyramids stone by stone on a strong foundation. Don't expect to reach your peak overnight. Add and test one item at a time before keeping something in your sales mix.

Little things add up

In summer 2008, *Inc.* magazine held a Best Lemonade Stand contest for young entrepreneurs. Here are two ideas you might want to steal for yourself.

Eleven-year-old Kaylee, from Massachusetts arranged a yard sale of gently used books and toys next to her lemonade stand. It kept customers browsing and buying longer.

Six-year-old Ethan from Arizona decorated his stand with surf boards as a real stand out. He put his stand on wheels so that he could take his stand to the customers during local festivals, parades or flea markets. What a way to "load the wagon."

Are there quality grass farmers near you who produce something you don't? Lamb. Cabrito. Eggs. Cheese. Be sure to label and acknowledge that these products come from like-minded producers.

Have you experimented with by-products? In addition to buffalo meat, Marlene Groves of Buffalo Groves sells buffalo tallow soap, painted buffalo jawbones, and buffalo skulls.

Don't overlook big bones sold for pet food. The average beef has nearly 50 pounds of bones as a processing byproduct. Even selling as low as 50 cents/pound can add $25 to a beeve. Veterinarians like to know of sources of "clean food" for their clients' animals.

What else can your farm supply? Herbs, dried and bottled with your private label. Flowers to grace the dinner table. Home-baked bread to serve with your cheese. Nuts or fruits from your trees and garden. Whenever it's appropriate, include recipes to eliminate the decision making for your time-crunched customers.

Selection. There's a fine line between offering too many choices and settling on the right number for your customer. This can only be determined by trial and error. Realize that given the opportunity to select by sizes, pounds, or items, makes it easier for your customer to buy from you.

Variety. One advantage over supermarkets of selling your grassfed products is the variety you can offer: goat cheese with chopped olives and garlic, lamb bacon, beef heart, smoked ham steak, bacon wrapped beef burgers or bacon wrapped lamb filets, and bones for pets. Anytime you have something new or newly re-packaged to offer, make a big deal out of it to attract people to your booth.

We belong to a vegetable co-op where everything is locally and seasonally grown. In winter when there are few offerings, I've been tempted to order things I've never eaten before. Once I ordered chard and had no clue what to do with it —eat it raw, cook it? How? Don't forget to include instructions to eliminate the guesswork for your customers.

Brainstorm around the centerpiece. It has taken me 12 years to decorate our second guestroom. Upstairs and out of mind, we rarely needed an overflow room. Safari-style sheets covering a mattress was good enough for the grandkids.

Then "real" guests from another country were expected, so I rushed around and found a bed frame and a baker's rack book shelf. For my tropical theme I moved in a wicker side table, a fake palm tree and a skinny, six-foot-tall, wooden giraffe.

Finally, I ran to Lowe's to buy a lamp. I must have jogged between the cash register and the lighting section five times—exchanging a chipped base for a better one, changing lampshades until a less pricey one rang up. The checker was very patient and helpful. Here's the point—just when I thought I had everything and was ready to swipe my credit card, she asked, "Do you have light bulbs?"

Ask yourself what your customers will need for one-stop shopping of a complete meal. Think beyond the food items. Cookbooks. Aprons—with your farm logo or slogan. Those of you with welder-loving spouses might even want to sell Argentine style *parrilla* grills. Anyone a potter making custom dinnerware, serving bowls and platters? Is that too pricey for your customers? How about sewing napkins and matching tablecloths or hand painting farm scenes or simple hearts and shapes on purchased picnic baskets?

With the greening of America, even Wal-mart is pushing the sale of reusable cloth shopping bags. Why not you too? Don't forget to have them printed with your logo.

Remember, you don't have to be like Henry Ford who made every single part of the Model T in his factory. He even kept a flock of sheep for wool used as stuffing in the seats. Instead, you can be a Toyota, which by comparison was assembled using parts manufactured by multiple companies.

Value added products

For many newcomers selling grassfed meats, one of the first problems they have to solve is how to sell all the ground meat that's left after the prime cuts are gone.

One solution is to turn it into something else by adding value to the raw product. Did you know that making a meat product ready-to-eat increases its value by a minimum of 500 percent It also allows you to control how it is cooked so you can guarantee tenderness and flavor.

In Allan's heritage food research he learned that few Americans had ever had marbled beef until the advent of the Shorthorn breed in the late 19th century. Once they tried it they really liked it. As it became more ubiquitous, beef replaced pork as the number one meat in the United States in most regions by 1900. However, because the Texas fever tick prevented the Shorthorn from being used in the Deep South, pork remained the meat of choice in that region until the mid 20th century.

Fat. Take a tip from the earlier era when people made lean beef more flavorful by combining it with pork or lamb fat. Today, tenderloin from even a well-finished animal is too lean to have much flavor, so wrap it in a strip of bacon and give it a fancy name. Let your customers know this is an excellent way to make tenderloin suitable for grilling. The pork fat not only adds flavor but keeps the beef from drying out while cooking.

Trimmings. With sausages, lean beef could be mixed with fat trimmings from other beef animals, pork or lamb.

Burgundy Pasture Beef packages "Ground Beef Trim" of heart, liver and kidney. Priced well below that of ground beef, it's aimed at people like us who want their pets to eat as healthily as they do.

Bones. The Yegerlehners in Clay City, Indiana, have experimented with creating Bone Broths from soup bones, neck bones, and feet. "It's hard to get much lower-end than that!" Kate Yegerlehner noted. She added that for centuries Bone Broths were revered for their healthful properties—mineral and nutrient rich gelatin. This is a value-added product because most consumers don't want to take the time to make it themselves. Just be sure your Bone Soup is made in a commercially-approved kitchen.

Hot, hot, hot. The Parrys of Fox Fire Farms in Colorado are finding goat and beef tamales a big hit at farmers' markets. They have them made up by a local restaurant and serve them hot with salsa. They have found that turning the entire goat into ground meat for the tamales each kid is worth $720. They found that legs, shanks and shoulder cuts are harder to sell unless someone wants to barbecue. Sometimes they save the loin and racks for individual sales.

What motivates purchases?

In a research study regarding the factors influencing buying decisions, reasons were ranked as follows:

1. Confidence.
2. Quality.
3. Service.
4. Selection

Price came in ninth.

Guerrilla Marketing Weapons, Jay Conrad Levinson

At our local market in Mississippi the tamale lady sells out in less than an hour.

Culture. In Germany, the majority of the beef eaten is consumed in sausage. For those of you in German descent regions, mixed meat sausages should probably be your primary heritage product.

In the Mediterranean region, the beef is too lean to hold together as a patty and so it is mixed with pork fat. Interestingly, some North American chefs now tout "Mediterranean style hamburgers" with pork fat and sell them for a premium price

In Greece and Turkey, beef and lamb fat trimmings are ground fine, combined with bread crumbs, water and oregano and pressed into cylinders in a hydraulic press. These meat cylinders are then slowly cooked on a rotisserie roaster. The meat is shaved off in thin strips with a large sharp knife and served in a pita sleeve with Feta cheese. These are called gyros in Greece and kebabs in Turkey.

In the Muslim areas of southern Spain, lean beef and lamb fat are coarsely ground similar to American hamburger. This meat combination is pressed into thin sheets and cooked similar to an American hamburger on a griddle. It is then rolled into a tube and served in a pita sleeve with Feta dressing similar to a gyro but it is called a kebab.

With the Muslim diaspora throughout Europe, kebab shops are becoming as common as McDonald's. They're our favorite for a quick meal in Europe. While many chefs have identified Spanish cuisine as the next "big thing," the biggest thing could be Spain's fast food—the kebab.

Overstock. David Evans of Marin Sun Farms in California got into value added meats out of necessity when a lot of roasts piled up. He turned them into pulled beef sandwiches, which he marketed at his booth at the San Francisco Ferry Terminal Farmers' Market. He spread out 18 different condiments for people to customize their sandwiches, and at times had 40 people in line. The pay off was a $2000 take home.

"Sausage is a good place to start," he said, "and jerky. It stores well and is not perishable if you can find a good processor. Salami is harder to do."

Repackaging. For a long time, Joel Salatin resisted his poultry customers' requests for cut-up chicken. Once he realized his customers would buy pre-cut chicken, or none at all, he bent to the pressure. Surprisingly, he discovered that he could earn as much for skinless chicken breasts as a whole chicken. As one who prices all farm tasks in terms of dollars per hour, cutting up chickens pays around $45 per hour.

Joel Salatin realized that many customers didn't know how to properly cut up a whole chicken, or didn't want to take the time to do it. Polyface Farm began doing the work for them. They now offer boneless, skinless breasts, chicken tenders, legs, thighs, and wings. They even package necks and backs in 5 lb bags for stock. That added $20,000 in revenue the first year —without growing any additional birds. He noted that customers are paying about $4 extra for "convenience" even though they can buy a whole chicken for the price of a boneless, skinless breast.

In Hempstead, Texas, the Crank family have learned that different customers want different sized birds. Most want a 3 to 3 1 /2 pound chicken. However, they often get requests for 2 and 3 pounders, and they have sold chickens from 1.6 pounds up to 5 pounds. They charge the same price per pound, giving their customers the service they ask for. They have also discovered an increasing demand for pre-packaged chicken parts. This allows them to use slightly damaged birds. At farmers' markets they always bring more product than they need so that they won't run out before the close of the day.

Festivals. Butch Tindell in Texas realized that his ranch initially didn't have the grass skills to produce a well-finished grassfed steak, so he concentrated initially on making deli sausages, processed meats, and hotdogs until the ranch's grass finishing skills got up to speed. This was an excellent strategy because what is an after-thought for most of us—the non-middle meats—became his main product. He has never found himself with an oversupply of hamburger as frequently happens to new beef marketers.

"A deli is a simple restaurant to run," he said. "A steak restaurant is more difficult." Selling deli beef at their Brazos de Dios community gave customers the opportunity to taste and see who they were. Beef products include Canadian bacon, pastrami, and corned beef.

Another excellent strategy for Butch and his Mennonite-dominant rural community has been an annual heritage food festival where they sell over 23,000 meals featuring chicken and grassfed meat products. There's no business like show business to add value to grassfed meats.

Service

If you and your neighbor offer the same quality products, how can you compete?

Add value to what you sell by adding service.

Think of it this way. Two stores have the same TV for sale. One offers a free service of delivery and set up. The other doesn't. Which would you choose, and why?

The first factor to consider is the reputation of the company. Have they been in business a long time? Will they be there if you need them later? If something goes wrong, will they back up what they sell by either fixing it or replacing it?

The second factor is service. How many of us are not electronic geniuses? Even if we are, having a hassle-free installation can tip the scales toward a sale.

Finally, what's the cost? That applies to the item itself and the service that comes with it. Many people—like me—would pay a little more for the business to do the labor and thinking for me. I don't have time to read an installation manual about something I know little about. I'd rather leave that to the "expert" who sells the product, can give me a quick lesson on how to use it, and can have it—and me—up and running in a matter of minutes.

Service for hire. Then there's the Service Contract. A very visible, highly advertised appliance store in our area offers Service Contracts. After one experience there where the Service

Contract (for a fee) was pushed almost more than the appliance itself, we never returned. This was just another way for the company to milk the customer. Their sales people were coached in how to make the customer feel guilty or worried if they didn't purchase the Service Contract. This Contract was an insurance policy in disguise with no real "service" about it.

Excellent service. "Service means solving your customers' problems, attending to their needs, making their lives better because they bought what you are selling," writes Jay Conrad Levinson, author of the *Guerrilla Marketing* books. He lists several common attributes of companies who excel in service.

1. They set amazingly high standards of performance. (Everything they do has the stamp of quality about it.)

2. They are obsessed with knowing what the customer wants. (They know that customers *are* their business. They solicit customer feedback. And they go the extra mile to give their customers what they want, even if it doesn't contribute to the sale. For example, perhaps you and your customer discuss a book about grassfed meats. The next time you see that person, loan them your copy of the book, or offer to buy one for them and collect their money later.)

3. They know that customer expectations must be understood and managed before they can be met and exceeded. (This takes listening, and selling the customer what they want and not just the thing you want to get rid of before you go home.)

4. Their products and services aim for customer satisfaction. (They look at everything as if they were the customer and ask, "What would I want?" Often this involves follow up to be sure they liked what you sold them.)

5. They're easy to buy from. (They take credit cards and Internet orders. And I'll add that they're not pushy or give you the hard sell if you decide to go away empty handed.)

6. They know money and attention to service will pay off in customer satisfaction and added profits for their business. (They look for ways to do more for their customers. They know

that the service they give can set them apart from their competitors.)

7. Everyone in their business is service oriented. (They are a company of team players. They're always looking for ways to do more for their customers with little or no added expense.)

Marketing mavens. These marketers want to help educate customers and prospects about grassfed products. They hand out brochures or flyers with every purchase and to those who are just curious about what they sell. When they find a good article about the grassfed industry, they display copies of it or alert their customers about books or articles in current issues of consumer magazines. They share what they've learned as a service. For free.

Veggie valet. San Francisco's Ferry Plaza Farmers' Market is huge and famous. It has an Information Booth staffed by volunteers for CUESA—Center for Urban Education about Sustainable Agriculture. The market's Veggie Valet allows

How to give good service

1. Figure out what the customer wants.
2. Get it for them—accurately, politely, enthusiastically.
3. Go the extra mile.
4. Make sure everyone who works with you knows this.

Bo Burlingham *Small Giants*

1. Make the customer feel as if they're unique.
2. Let the customer know you want to be of service.
3. Stay in contact with follow up.
4. Exceed the customer's expectations

Jay Conrad Levinson *The Way of the Guerrilla*

shoppers to drop off their purchases while they continue shopping among the vendors. They can also hold all the purchases until the customer can drive closer for curbside delivery.

If your farmers' market is too small for either of these services, offer them yourself. Be sure to post signs to encourage people to take advantage of these services since they can make you stand out.

In *Meatball Sundae*, Seth Godin writes that businesses organized around service will be the winners in this high-tech age. "There isn't room in the marketplace for every competitor to do this, and the first major provider that does establishes a bar that the others rarely reach. As a result, word spreads, and competitive insulation results."

Celebrating special occasions and milestones

Special occasions—Christmas, Mother's or Father's Day, Fourth of July—offer great opportunities to celebrate special purchases of your grassfed products. However, to standout from your competition, create your own celebrations.

Gift baskets. Gift baskets serve two purposes. They solve a problem for your customer who may be stumped on what to give for that special occasion. And they allow you to sell multiple products in one fell swoop.

Offer two or three price levels of gift baskets. Budget conscious shoppers may want just the bare bones basic gift package, which should contain the least expensive items. You might find colorful wire baskets or plastic beach buckets to hold everything. An average priced gift package can be assembled with more side items or pricier containers such as a pretty basket or box. The ultimate gift package should "wow" potential customers—pack everything into a completely outfitted wicker picnic basket or include such things as embroidered napkins. What you put inside is limited only by your imagination. With each purchase level include a fill-in-the-blank gift certificate for your grassfed meat or cheese products.

Assemble a small booklet of recipes (or include a cookbook), spices to use in the recipes, and gift coupons with blanks to fill in any amount (be sure to include your logo and contact information in the design) for your perishable products.

Depending on the season, tuck in a tiny potted herb or two.

Offer gift baskets for different occasions—a picnic getaway (include red checked napkins), anniversary (drop in two inexpensive fluted glasses), birthday (a box of candles). Include a greeting card to save your customer the time of shopping for one. Have several cards on hand to cover birthdays, anniversaries, congratulations, or thank yous.

Gift bags. You can use any kind of bag to assemble your gift assortment. Google promotional products or gift specialties to find companies that offer bulk orders of low cost items and can imprint your logo on them.

Boxed orders. Follow Joel Salatin's lead. He assembles various cuts of meat—all beef or all pork—boxes them into packages of the same weight, and gives them fun names—Butcher's Choice, Slow Cooker's Dream, Meatloaf Mania, to name a few. This takes the guesswork out of "What should I order?"

Turkey talk. If you sell turkeys for the holidays, try this to get customers talking (and buying). Deliver it in a brown grocery sack (with your logo attached). And include this recipe for cooking a turkey in a bag:

1. A day or two before cooking assemble the marinade. Cut slits in the skin working from the back forward and tuck spoonfuls of marinade beneath the skin.

2. I use a variation of this marinade, so you can easily create your own. 1 cup olive oil, 1 tablespoon poultry seasoning (I use a Cajun seasoning mix), 1.5 tb. salt, half a cup of chopped scallions or onions, half t. black pepper, 2 tb. paprika, half of a t. cayenne, 2 tb. minced garlic. Wrap turkey in foil or plastic wrap.

3. The evening before the event, butter the inside of the brown paper sack. Place the turkey inside breast side down. Close the bag with staples. Place in a turkey pan.

4. Cook at 450 degrees F for 1.5 hours. Be sure the bag is positioned in the oven without touching the sides or elements.

5. Turn the oven off and leave the turkey alone for at least 4 - 5 hours.

I've used this recipe three times. You will smell the bag when the turkey is cooking at high temps. We try to put this on as late as possible so that we can discard the sack and separate the juices early in the morning. The turkey comes out browned and super tender.

Milestones. These are memorable events. People spend weeks, months, sometimes years planning a major milestone. Done well, guests will talk about the event long after it's over. That's the best free publicity anyone can get.

Think about the milestones you can market to your customers:

First day of school. Mothers often get so anxious and emotional thinking about their child's first day at school they take little time for anything else. Ease that separation anxiety with suggestions for that momentous first day. "Pack a lunch" menus might include cheese sandwiches, cold brisket bites, homemade cookies (from your farm), a piece of fresh fruit. That's also a wholesome bagged lunch for over-worked office workers.

Your date's first home-cooked meal. Everyone who's in a new relationship is nervous, even if this might be the second, or third date. Minimize the fluster over what to fix for a first meal at home with these ideas. Keep it simple so that any

True milestones

Milestones were granite obelisks used by the Romans on the Appian Way to mark distances from a central stone set in the Forum by Augustus, Rome's first emperor. The name came from the Latin *mille*, which means "thousand." A Roman mile was measured by 1000 paces, each of which was five feet.

nervousness about the date won't affect the cooking. A special cut of meat or hearty steak will win over any guy. It's also nice to hand over the spatula and ask him to do the grilling. This inspires confidence in him and allows him to play with fire, something all guys love.

For my first meal, I made a cheese fondue for Allan, a true meat-and-potatoes guy. Even though he went home hungry, he married me anyway. However, the next time he invited me out to eat rather than trust me to cook something that underwhelmed him.

Weddings. To begin, you may want to sell a local caterer on using your better-than-ordinary beef, lamb, or chicken for their menus. Some grass farmers have become so well known for their delicious and healthy products that being a caterer becomes a spinoff enterprise. A gorgeous pastoral background is a natural choice for a shower, reception or the wedding itself. The next step is to build a stand-alone reception or meeting area, which could lead to your farm being a small conference provider.

Mini-milestones. Celebrations don't always have to be grand events. Women mark small milestones too. How about a bride's first meal. Hopefully she is better clued in by then as to what her husband really likes.

Mark the move with a first meal in a new home or apartment.

"From Hamburger to Steak" could celebrate a first or new job. Similarly, getting a raise or new position is cause to dine well on pricier cuts.

"We've Retired" sampler packages might include special offer coupons to reward your best customers.

Every occasion can serve as an excuse for a marketing milestone—buying or selling a home or business, sending a child to college, even new beginnings after a divorce. Create event reminders to give your customers reasons to buy.

Celebrate yourself. Customers aren't the only ones to celebrate. Include your customers in your own milestones—the anniversary of the year you changed to 100% grassfed, the birthday of your family's farm, you've just added pork to your enterprises, you've expanded with more leased acres, you've added a new employee, your cows or lambs have just given birth. Let people know how long you've been in business. Anything that's old, new or different is worth celebrating.

Beyond farmers' markets: the farm store

There will come a time when your production is ramped up to a level where you have more product than you can easily sell at farmers' markets. While they're great places to get a leg up on the meat-marketing ladder, farmers' markets should only supply a portion of your total direct market sales.

It's time to diversify. Where do you go next?

On-farm store. The most obvious solution is to build an on-farm store. It's close to home, which can be a real time saver. It's the ultimate low-cost distribution system to have your customers come to you. You can set opening hours to suit your lifestyle—Saturdays only, or 10:00 am to 2:00 pm weekdays, whatever works for you.

I suggest building rather than converting an existing building, simply because those who have done so have learned the hard way that it can be cheaper to build rather than renovate. Plus, you won't be boxed in by existing walls or features of the older structure. You can create your dream store, and build to allow for future expansion.

No longer subject to the rules of some farmers' markets (no pets, no crafts, no theatrics), you can customize your store to reflect your personality. (Do check local building codes and restrictions regarding the sale of food items.) Joel Salatin keeps his on-farm store purposely plain and nostalgic with a wood-burning stove and hand-written invoices. When it was only open a few hours on Saturday it served some 400 families within a 75-mile radius. More recently his remote-phone-call employee moved with the phone to the store allowing it to be open more often.

Pick a theme—country, antiques, Native American, sports, postcard travels—and go wild decorating. If we had a farm store it would surely have a train theme for Allan. (I'd love it just to move some of his model trains, books and magazines out of the house.)

Located just off a busy four-lane highway that connects Anniston and Gadsden, Alabama, David and Leighann Wright have a tiny 200 square foot on-farm store that rings up nearly half a million dollars worth of retail sales each year.

Open six days a week, the store sells 2500 gallons of milk a week, plus Wright's farmstead butter and ice cream, and grassfed beef and eggs from another Alabama grass farmer (who in turns sells Wright's ice cream). Retail customers can watch the milk bottling and ice cream manufacturing through large glass windows from the retail area.

Wright uses a small batch, hot water, pasteurization method that preserves most of the milk's natural enzymes and flavor. "Our milk tastes much more like raw milk than commodity milk. We call it 'Old Fashioned Flavor,'" David said.

The milk is also sold unhomogenized and the dairy touts its milk as "where the cream still rises to the top." David said it is this superior flavor that encourages people to drive considerable distances to the farm to get his milk. He added a drive-through window for his busy customers' convenience.

Dairy to go. David Wright said the major problem with an all-fluid milk product is that it is not storable long-term. This makes matching supply and demand fluctuations a real challenge, particularly for a startup, growing operation. He initially solved this problem by selling all of his milk to the local dairy processor and then buying back the portion he could retail. Of course, the retail portion never actually left the farm and was just a paper transaction.

While this allowed him to easily market all of his milk, David admitted that the processor's "money for nothing" grated on his nerves. He began looking for a storable, manufactured product he could make. His solution was cream. "Cream can be frozen and used whenever," he said.

Today, he uses his cream to make farmstead butter and ice cream, which eventually enabled him to terminate the buy-back agreement. To hasten that day of independence, he started wholesaling milk and cream. His largest milk buyer is a full-line restaurant jobber.

"We discovered that restaurants are an under-served milk market. The big guys won't stop anywhere for less than $200, but my jobber is already going there so adding milk was a natural for him."

Another wholesaler sells into upscale Birmingham specialty grocery stores. "It took my breath away when I saw my milk priced at double what I charge in my own store."

To tap into this high dollar urban market, David "franchised" a Wright Creamery ice cream parlor in Birmingham to his niece. To further increase ice cream sales, he placed 20 small freezer boxes in small grocery outlets within 30 miles of his farm. These boxes are serviced by a small delivery truck once a week.

Details. Have bright lighting inside. This shows off your products better—shows you have nothing to hide—and is a convenience for those seniors who don't see as well as us younger whipper snappers. Use spotlights for key items.

Hours of operation. Research has shown that you can increase sales just by increasing the time you are open. If you don't want to be behind the cash register the whole time (and you shouldn't be), hire someone. Know the patterns of your customers. Are they working moms? Match your store hours to their commuting times—opening 4-7pm. Are they soccer moms? Open on Saturdays. Offer Hoot Owl shopping with one late night closing—Thursdays are good as a set up for the coming weekend. What about Sunday afternoons? Encourage your customers to bring their kids and make it a family outing. Be sure to advertise your hours on your website, brochures, flyers, order blanks.

If you take Internet orders, be sure to respond no later than the following morning, or have an automated system to immediately acknowledge an order. Use your computer or fax to quickly get information to prospects and customers.

Keep them there. Just because you get the customer in the store doesn't mean you can stop marketing. Use in-store signs to point out benefits, something special, new, nutritional facts, or post Did-You-Know? questions and answers with facts and figures throughout the merchandising areas. Post them anywhere someone is inclined to wait.

Why bother with farmers' markets?

Farmers' markets are not and should not be your only marketing outlet. In fact, there may come a day when you decide it's easier to market your products without them.

They can be exhausting and time consuming. They have rules and regulations.

However, they offer an important service. They help you build a solid customer base and customer loyalty.

Once you own your customer, you want to be able to sell them more products more often. These are the people who may begin buying a single cut of meat and graduate to buying a quarter animal, or a half, or—success—a whole, especially when you point out that buying quarters or halves cost less per pound than buying individual cuts.

Always aim for ways to entice your customer to buy more.

Keep your customers informed and entertained. The materials you display can let them know you're staying abreast of grassfed happenings, and that will encourage them to always come to you for your knowledge and industry leadership. Don't let those in-store messages collect cobwebs. Keep them fresh. Rotate them regularly.

Mix it up. Change displays frequently. Make them inviting, innovative, seasonal, and to call attention to specials. Decorate your windows to draw people inside.

Expanding the farm store. Elsewhere, we've discussed "loading the wagon" with products from other quality grassfed producers or complementary items. Consider selling food or crafts unique to your state. Check with the Economic Development Department in your state capital to learn about marketing fairs. This will give you the opportunity to meet state-wide vendors who are only too happy to piggy back their sales with yours. Assemble these items into gift packages—barbecue sauces, condiments, mixes, or spices—to "beef up" your meat sales. Don't forget books by local authors as well as cookbooks.

To help sell their excess hamburger, Wendy and Jon Taggart in tiny Grandview, Texas, put a commercial kitchen and dining area in their off-farm store—located four miles from their farm. They use two picnic tables for seating inside and another couple outside. With no advertising but a sign outside, they sell around 100 hamburgers a day, priced about five times higher than typical fast food burgers.

This cooked-to-order gourmet hamburger encourages customers to buy their beef and other products at the Taggarts' Burgundy Pasture Beef store. Lighted, glass-fronted coolers surround the picnic tables. Wendy keeps free recipe sheets in a bin with dividers that look like a nesting box for chickens. Now, thanks to years of home deliveries, they know where their clients are clustered in the Dallas-Fort Worth metroplex and are thinking about opening a second store in an urban location.

Outdoor signs. Throughout the South highway travelers can see red bird houses with black roofs and barn roofs and siding painted with signs that read, "See Rock City." Some barn roofs expand the message to include "See 7 States." They're all invitations to visit Lookout Mountain, Tennessee. Today, there's a website where you can purchase a See Rock City birdhouse.

If you're going to use billboard-size signs keep the words to a minimum and make the letters big enough to be read quickly and easily. A six-word message is best. Would your farm's slogan work?

Wall Drug in Wall, South Dakota, planted signs every few miles advertising Free Ice Water.

The second most important thing to include is directions, short directions: Turn here. Next exit. 5 miles ahead.

Use the side of the barn if it's near the road. Think visually. Think purple cows. With everything you do look for ways to be different.

Location

Location, location, location. Yes, it matters.

You want things to be convenient for your customers to buy.

Case in point. Walgreens' drugstores made break-through sales when it positioned itself to be the best at convenience for their customers. They looked for corner locations to make it easy for their customers to enter and exit from different directions. According to Jim Collins, writing in *Good to Great, Why some companies make the leap...and others don't*, they would close down a store and move if a better location became available. In major cities, they clustered stores within a small radius and pioneered drive-through pharmacies.

When considering convenience, it doesn't hurt to think about convenience for yourself. For some producers, having the farm store off the farm eliminates interruptions while they're involved with vital pasture improvements and activities. That doesn't mean no one is ever allowed on the farm. Customers want transparency to know where their food comes from. It just means you choose the times—a periodic field day or open house—when those customers are most welcome to look over your shoulder.

Enroute. David Evans of Marin Sun Farms in the greater San Francisco Bay Area is one who has his butcher shop and deli off farm. Due to zoning the only way he could have a butcher shop was if it was attached to a restaurant. He opened a small 25 seat restaurant called "Marin Sun Farms Eatery" in order to control the quality of the meat cutting. "Cutting your own meat is the only way you can control its quality. My customers are very particular about how a piece of meat looks on the plate.

"We were primarily a retail meat shop, but for $10 we would cook your steak and serve it with locally grown organic vegetables," he said. "A hot product for us is 'natural beef tallow french fries.'"

The restaurant was a marginal enterprise but worth it to get the butcher shop.

Then disaster struck on New Year's Eve 2005. The small river adjacent to his butcher shop flooded his restaurant/ butcher shop.

"I thought it was the end of the world, but in retrospect, it probably saved my life. I was growing entirely too fast. I didn't have time to think. After the flood I had plenty of time to think." Luckily, Evans had flood insurance, and the flood resulted in a dramatic downsizing.

Marin Sun Farms Eatery is located on Highway 1 enroute to popular Point Reyes National Seashore. "If you're not the destination," he said, "then you have to figure out how to get people to stop on the way to their destination."

Although he has outdoor seating at the Eatery, zoning won't allow him to barbecue outside. "That would be ideal for us," he said, since there's a traffic stop just up the road from the eatery. He feels just the smell of barbecuing meat would entice people to the Eatery. "You could burn sausages just to get the smell," he said.

Instead he is experimenting with different signage to engage weekenders to pick up their picnic supplies enroute to Point Reyes and then stop again for meat products on their way home. Fortunately, traffic at the Eatery is picking up because he has been in that location long enough for name recognition.

Parking. Ample parking is critical, however, it's not a

killer if the location is otherwise perfect, such as a place that gets a lot of foot traffic. Zingerman's Deli in Ann Arbor, Michigan, started off with and still has limited parking, yet their remarkable products have allowed them to grow into a multi-million dollar business.

Do you have a responsible teen who could double as a valet?

Drive through. "Soccer moms don't have time to cook," said Buckingham, Viriginia, direct marketer Terry Hudgins. "Newlyweds don't know how to cook. Anytime you can cook something so that it's ready to eat, that's value added." And priceless to the consumer. The drive through window makes grassfed convenience food.

Taking a cue from Walgreens, customers filled out orders online and then picked up their prescriptions at the drive-through window or had it delivered at home.

Easy to find. Don't let cheap rent drive the location you choose for an off-farm store. However, customers will drive to remote locations for your quality products. Have maps printed up and include with every purchase. Post it on your website. Add it to your catalog. Give street by street or highway directions and also include landmarks along the way. Although we

Questions to ask yourself when choosing a location

1. Will your prospects be walking or driving past the store?
2. Will they come to you for their purchases?
3. Are there any restrictions on signage?
4. If you're leasing, are you allowed to make physical changes?

Remember, people will drive out of their way for great values.

Guerrilla Marketing Weapons, Jay Conrad Levinson

have been there numerous times, we always need a map to get to the Salatins' Polyface Farm. Plenty of his customers make the trip to his farm like mice in a maze following the lure of cheese.

Home delivery. Absolutely the most convenient means of buying their food is home delivery. That's where Buyers' Clubs are unsurpassed for customer satisfaction. With pre-ordered and guaranteed payment upon delivery, the producer can match supply to demand and sleep better at night knowing all sales are final.

Your challenge is how to get your products from grass to plate in the most direct and convenient manner for you and your customer.

Organizing a Buying Club

Joel Salatin is an innovative, imaginative producer and marketer who lives in the Shenandoah Valley in the middle of nowhere, otherwise known as Swoope, Virginia. For a number of years he and his family sold their Polyface Farm products — eggs, chicken, beef, pork and more—at the local farmers' market. However, his favorite—and easiest—form of sales has always been direct marketing from the farm. Since he believes in selling locally, he doesn't offer shipping services.

Several years ago a group of his female customers asked what it would take for him to deliver his grassfed products to her and her friends who live in the greater Washington D.C. area.

Let's take a closer look at what happened and how this works.

Make it worth your while. A key point is that Joel's customers asked for this. He didn't wake up one morning and decide to form what he calls a Metropolitan (rhymes with cosmopolitan) Buying Club. He made some calculations and decided if she could guarantee a minimum order of $3000, he would be willing to make monthly deliveries.

This single club has grown into some dozen Buying Clubs and accounts for a little over half of all of Polyface Farms' sales. Thanks to Joel's daughter-in-law, Sheri, the farm marketing director, their Buying Club families have exploded.

Joel said it took three years to go from 30 to 60 families, two more years to reach 120 families, but now, thanks to Sheri, they sell to 800 families in a single year. And, the Metro Club buyers spend more than those customers who buy directly from the farm. New customers are surprised to learn there is no fee to be a member of a Buying Club.

Each club must average $1000 per delivery, and new groups are given a year to achieve that goal. Only two purchases a year keeps each Club in the "active" files. There are eight drops, each averaging $8000 annually.

The selling advantage over farmers' markets is that everything is pre-sold. At the end of the day the Salatins don't return to the farm with anything. As a result, they no longer sell at farmers' markets.

Copy the model. Intrigued by the Salatins' success, David Evans of Marin Sun Farms in California emailed his farmers' market customers suggesting a six-month Community Supported Agriculture (CSA) subscription. About 150 customers readily signed up.

A long-time student of Joel Salatin's production and marketing approaches, he said he saw a lot of advantages in Joel's Metropolitan Buying Clubs over farmers' markets. He was also intrigued by the many vegetable CSAs (Community Supported Agriculture) in the Bay Area and wondered if the two models couldn't be merged.

With a CSA, the consumer buys a share of the farmer's crop and pays for it in advance. A major problem with selling meat by the cut is that consumers tend to cherry pick the tenderloins, ribeyes and other middle meats and avoid the hamburger and roasts. In a vegetable CSA, the consumer's share is a variety of what is available at that time and the farmer chooses what the consumer will receive.

Evans decided to try a CSA-style farmer's choice marketing model with meats.

The Marin Sun CSA offers a six-month subscription. "I wanted the initial subscription to be long enough for people to get in the habit of receiving their meat this way before they faced the renewal decision," David said.

There are several package options. These range from a $1350 all steak package to a $195 all-sausage package. The most popular package was one with six pounds of steak, ten pounds of roast and eight pounds of ground meat.

All of the meat is sold frozen and typically includes some meat from each species—beef, lamb, goat and pork. Consumers are allowed to opt out of pork or goat. "Most of our customers are sophisticated foodies and enjoy receiving a variety of meats. People here are enthusiastic about expanding their culinary talents."

Marin Sun has seven pickup points scattered over the Bay Area where the customer can pick up their weekly pre-package portion of meat. The benefit to the consumer is that they can save 15 to 20% over the price of buying the meat at the farmers' market. They also receive a 5% discount on any additional meat they buy at our retail butcher shop and can buy bones and offal for their pets.

He has been picking up an additional five to ten customers each week through word-of-mouth. The benefits to Marin Sun were that they get paid in advance, can sell the meat frozen and he can choose what meat goes in the package. "This really helps balance our sales," David said.

Transportation. In the early stages, Joel bought at an auction a used stretch van, which had served as a prisoner transport. It could accommodate several large coolers organized and packed at the farm with everyone's order. (All meats are sold frozen.) He rigged a rooftop air conditioner unit so that the van could stay cool without running the engine while deliveries were unloaded. During seasonal bulk meat buying periods a U-Haul trailer was pulled behind.

For greater efficiency, he eventually sought out a used reefer truck.

Delivery fees. Catalog shoppers are used to paying freight and handling fees. Other farmers who routinely ship products by Fed Ex pass shipping fees on to the customer. Adding delivery fees to cover fuel and labor is the price of doing business. Joel charges 25 cents per pound. After all, he's saving the customer time and energy they would otherwise spend driving the 300-mile round trip to the farm.

One tip he uses to keep the delivery charges from appearing too high is that he never offers two heavy items on the same day. For example, he never offers holiday turkeys and beef quarters on the same delivery day.

Prices also look better by keeping the delivery fees separate rather than building the fees into the price of the products.

This also rewards those who buy direct from the farm with lower prices.

From an accounting standpoint, the delivery fees can be categorized to track transportation profitability.

Joel Huesby of Thundering Hooves Farm adds a 7% delivery fee to customers' bills.

Scheduling. Sheri Salatin organizes the orders by email and sends an email "newsletter" reminder a week before delivery. The timing is critical. Two weeks is too far ahead and less

Advantages of Metro Buying Clubs vs farmers' markets

- Everything being pre-sold eliminates speculation.
- Guaranteed sales.
- Eliminates guesswork about volume of sales products.
- No inspectors.
- No commission fees.
- No market rules.
- Nothing to bring home after delivery.

Joel Salatin

than a week is too late to optimize sales.

The newsletter tells the time of delivery, available seasonal specials, Polyface Farm news, has an occasional recipe, and lists the total amount purchased by all Buying Clubs on the previous month's delivery. Publishing the amounts creates a sense of rivalry among the Buying Clubs, which indirectly increases sales.

Orders are taken on a first-come, first-served basis, with the time and date of the order written when it is received. This helps maintain harmony among customers when a particular item sells out.

Joel said to never give a range of times you will be there. Instead, particularly when there will be multiple drops at one location, be specific. For instance, A-L will be at 10:00; M-P at 10:30; Q-Z at 11:00.

Deliveries are made every six to eight weeks. This is about the length of time eggs can be stored, and it gives time for the volume to build up.

The drop site. A pre-arranged drop site is agreed upon. This is generally at the home of one of the ladies, the hostess, who takes responsibility for distributing the orders and collecting the money for the customers in her Buying Club network. Joel prefers private residences rather than public places.

On the other side of the country in Washington state Joel Huesby likes the term Neighborhood Buying Club for his buyers. "We want to meet people where they are," he said. They also have Business Buying Clubs with drop sites at a specific company building.

In Joel Salatin's backyard, Mark Lilley, an enterprising entrepreneur who is not a farmer bought an old school bus and had it painted as if by children with stick figures and hand prints. He scrawled his business name "Farm to Family" across the sides. Inside he pulled out the seats and replaced them with chicken-wire vegetable bins. The back of the bus has two rocking chairs and a miniature pot bellied stove for an old-fashioned country store decor. He rounds up seasonal items from Joel and other farmers and drives them to office complexes in Richmond, Virginia, after sending them a well-publicized schedule. People take a short break from work, shop, and then return to the office creating a workplace buzz and new customers. What impressed Joel the most was that in spite of selling to two of his own Metropolitan Buying Clubs in the city, thanks to Mark, Polyface Farms picked up 75 additional customers within two weeks.

Team deliveries. The Salatin newsletter doesn't give directions to the delivery home. The Salatins have found that hostesses are better at giving their new customers directions to their home.

A crucial element to their success is to always arrive on time. Customers who are late or don't pick up an order are dropped.

Two people always make the deliveries so that one person can spend time chatting with the customer while the other person watches the clock to make sure they stay on schedule. A typical drop will take half an hour to unload and collect payment.

Lagniappe. Customers who bring new customers into their circle are always rewarded with a little something extra. Other unexpected gifts for high-producing hostesses might be zucchini bread baked by Joel's daughter, Rachel, or perhaps some extra eggs. The best hostesses and those who order every time are rewarded with pickled eggs at Easter, pumpkins in the fall, cinnamon rolls at Christmas.

Anyone who brings in a new customer gets a $10 credit or $10 deducted from their bill. Rewards stimulate word of mouth.

For a little extra advertising, zip-lock bags are used to organize items in the cooler. These bags say "Healing the Land One Bite at a Time" and have the Polyface logo on them.

Load the wagon. In addition to all of Polyface's offerings, the Salatins work with other farmers in their area to seasonally sell as many additional products as possible. Some of these products include organic produce, artisanal cheese, homemade salsa, and organic apple juice. One farmer drives two and half hours to take his organic vegetables to Polyface for delivery day.

Joel Salatin's payment for loading the wagon with non-Polyface products is the delivery charges generated by the extra items from other farmers.

Find a way. During a producer panel discussion at a *Stockman Grass Farmer* conference, the question was asked, "If you were starting over, what would you do differently?" Overwhelming everyone agreed that they would find a way to start a Metro Buying Club sooner.

Are Metro Buying Clubs in your farm's future? You do the math.

Beyond the farm store

Your store, on or off the farm, is booming, and your production capability can allow you to sell your products elsewhere. What's next?

Restaurants. Avoid restaurant chains. Often, the corporate structure is such that the chef is locked into buying from mega-brand food distributors.

As an example, Joel Salatin tells the story of a Virginia woman whose catering business had evolved into a medium-upscale restaurant. While dining at her restaurant one evening, the Salatins were joined by the owner. She explained how all the national franchises were competing for her customers. Their recognized names and heavy capital was impossible to match, but there was one area where she could set herself apart from the crowd—that was by establishing her restaurant as a venue for fresh, local, seasonal ingredients. She was already familiar with Polyface Farm products, having used the Salatins' chickens for her catering clients. Now, the Salatins are supplying grassfed Angus beef and Polyface pork.

David Evans of Marin Sun Farms near San Francisco suggests targeting famous-chef restaurants in order to be noticed. White table cloth restaurants represent 50% of his sales and approximately 75% of his total volume.

It's also best to avoid restaurants that serve ribeyes and strips. You want to sell to one that features roasts, briskets, and hamburger. Remember, regardless of weight, a steer only has 24 ribeyes and 24 New York strips in it, so don't put yourself in a situation where you're limited by individual cut sales.

Rolling restaurant. North Carolinian Randy Sipe purchased a panel van with a side flip up shade for a rolling restaurant. He takes his Rawhide Rancher to agricultural events, Oktoberfests, bluegrass festivals, antique tractor rallies and other outdoor events. The van has a plastic wrap sign such as those used on buses or buildings. The main picture is of Black Angus cows grazing green pastures with inserts of grilled beef. A list of grassfed benefits, his website and phone number are large enough to be read while on the road. He sells cooked burgers and country fried steak. Because of his unique product and colorful vehicle, festival managers always find room for him to set up shop.

In-home chefs. A hot luxury for urbanites is an in-home chef. These roving chefs shop for all the ingredients for a meal, then cook and serve it in the client's home. If entertainment cooking is second nature to you, this could become a lucrative sideline.

Cooking classes. If your town is so small that the only restaurants are fast food chains, look for area caterers. Team up with instructors of Gourmet Cooking classes. Seek out delis, tea rooms, the Country Club, and locally owned cafes. Don't forget wedding planners.

Promote and create your own cooking classes using your farm's products in kitchen stores. If your farm store doesn't have a commercial kitchen, check with community colleges or universities that offer Lifelong Learning classes for leisure-class adults. You might even suggest summer school cooking classes for children. My friend Patricia's granddaughter gave her the idea to create her thriving Kids in My Kitchen business.

Grocery stores. The same thing applies to food stores. Seek out the little guys who are desperate for ways to compete with the giants. "Local" has cache in today's food world.

Wouldn't anyone prefer a locally-grown, vine ripened tomato to one that's machine harvested, hard as a tennis ball, and shipped hundreds or thousands of miles to market?

Natural food markets and health food stores would be the easiest to approach since most are already sold on healthy lifestyles.

By offering more variety, you will be getting the edge up on the big box stores that are trimming back the variety of brands and goods they sell. This is the time for individual direct marketers to become the "go to" place for fresh and healthy food.

Be civic minded. Nearly every civic organization has a need for food events, even if it's just an annual holiday celebration. Get in touch with AARP, Rotary, Toastmasters, or business group breakfasts. If you can piggyback a talk about the benefits of grassfed products with serving that first meal, you open a window for potential new customers.

Churches. Most churches provide several food events throughout the year for their congregation. Our church's Sunday School classes have seven food events annually, and that's in addition to monthly luncheons to welcome newcomers and seniors, festivals and special events. Offer to be their meat supplier. You might also tip the scale for this kind of sale by offering to direct the church's cooking team for the first event.

Hospitals. Approach smaller community hospitals first, not only because they may be more open to products from local vendors, but also because their volume needs will be smaller and more manageable. Maybe your initial sales will be to the hospital's cafeteria rather than all of the patients under their care. Explore assisted living facilities and retirement communities.

Target a smaller group within the health sector—cardio or weight-loss clinics, for instance. Do you have local health clubs that serve food? Be prepared to promote and defend your product, not only verbally, but with written material if you are working with a staff nutritionist. Many are still clueless about grassfed benefits. It hasn't yet become part of the mainstream curriculum.

Schools. Schools across the nation are on a race to eliminate junk food in vending machines and for meals. With the rising rate of childhood obesity, now is a good time to promote healthy, grassfed products. A desire for healthy alternatives can extend to school fundraisers. Green and wholesome are today's buzz words. Parents and teachers are getting away from products that aren't made in the USA. They want something other than cookie doughs, donuts, or candy bars to sell for fundraisers. Suggest products from your farm—handmade soaps for example. After all, it's the children's mothers who are the purchasers, and they're more likely to buy something they want for themselves.

Universities. Hardwick Beef, a grassfed beef company based in Hardwick, Massachusetts, was able to get its beef into Dartmouth College by taking the issue directly to the students. Tiring of the delay in dealing with college bureaucrats, Hardwick set up a tasting table on campus and passed out free samples.

The college's bureaucrats retaliated by posting a big sign near the table that said, "The regular burger you get costs $4.50 and if we bring in this 100% grassfed beef it will cost $6.85 so we want you to vote."

They did.

Out of a total of 92 votes cast, Hardwick grassfed got 87.

Here's another unobtrusive way to get your grassfed products into college. Talk with their theater director and suggest menus for Dinner Theater.

Although he spreads his sales among farmers' markets, and restaurants, David Evans wholesales hamburger, chuck roasts and bottom rounds to Stanford University. "Big wholesale customers are scary," he said. "I much prefer having lots of small accounts to a few big ones, because if you lose one it doesn't hurt too much."

Evans warns that many wholesale companies don't pay for 90 days. This means you need a large cash cushion to stay afloat between payments. "I believe you have to have some high margin retail sales to be able to afford wholesaling. You certainly can't build a small meat business on wholesale sales alone."

Urban farmers. Here's an idea Joel Salatin ran across during one of his many speaking engagements. During the economic downturn as more urban families turned to backyard vegetable farming, often with underwhelming success, some farmers thinking outside of the box set up urban farm services for these city folks. Think of it in terms of the lawn care services already proliferating throughout the USA. These farmers charged on a square foot basis and set up regular times to work in the backyard gardens—just the way the lawn boy regularly comes by for mowing. This could be a great enterprise for farm teens who are of driving age.

Virtual market. Local Dirt is a website founded by former Whole Foods employee, Heather Hileren. Based in Madison, Wisconsin, the website links local farmers with food buyers, restaurants, and grocery stores. People who order from the site learn the source for traceability, and the distance the item traveled from the farm to the purchaser. Businesses or farms that place orders pay an annual subscription fee. She tracks quality by having restaurants rate deliveries.

Outstanding in the Field. Jim Denevan, who calls himself a chef, artist and surfer, founded Outstanding in the Field in 1999 to honor farmers and their work. Traveling around the country in a 1953 red and white bus, Denevan and his team have hosted food lovers on farms and gardens amidst the crops and flowers, on a mountain top, in a sea cave, refurbished barn and a green house, which doubled as a rain event backup. His mission is "to reconnect diners to the land and the origins of their food, and to honor the local farmers and food artisans who cultivate it."

Participants make online reservations and pay an average of $200 per person for a five course meal. "The dinners always take place between the soil and the sky," according to his website, "near the source of the ingredients for your meal." After all, it isn't every day that urbanites get to sit next to the person who raised the animals or grew the vegetables that are served farm-style at long, tablecloth-covered tables. This roving culinary adventure has even taken place in Tuscany, Italy, and is described as a restaurant without walls. In some cases they ask participants to bring their own plates. This gives the white tablecloth tables a festive appearance, and cuts down on the cost of renting dishes.

Katama Farms in Edgartown, Massachusetts, hosts an annual summer fundraising dinner and auction for The Farm Institute called Meals in the Meadow. The Institute is a teaching farm that educates and engages children and adults in sustainable agriculture through the diverse operations of a working farm. The buffet meal is catered by area restaurants. Music and dancing follows the auction which includes this item, "You can have laying hens at your home for a week!" In 2010 seats were $225 per person; Shepherds received two tickets for $1000 donation and Harvesters received 6 tickets for $5000.

Fox Fire Farms in Ignatio, Colorado, held their first

Are you amused yet?

Ask yourself where do most children go after school? The bowling alley, skating rink, go-cart track? If these places sell burgers, suggest they grill grassfed.

Plan to have a booth at the next State Fair or local festival.

Ideas are all around you on where to sell grassfed products. Just follow the scent of cooking.

Dinner at the Farm the same year. Each course featured products grown at the farm and prepared by chefs from three area restaurants. Wine pairings were set for all five courses. The cost was $100 per person.

Yes, it does take a lot of work, but what party doesn't? Most importantly, each event begins with a tour before the meal. And you can bet it's a memory talked about long after the last guest has found their way—in the dark—back to her car.

Convincing chefs to go grassfed

Having a restaurant, particularly a prestigious one use and promote your grassfed products can be a coup. First it provides an excellent public relations tool by having an outside entity endorse your products. And it's a win-win proposition for them because you will surely promote their chef and restaurant in your marketing venues.

How do you get them onboard in the first place?

Rule number one. When making an appointment to meet, make sure you talk to the chef and not a cook. Chefs are the ones who bring prestige to a restaurant. Cooks are the ones who follow orders set by the owner or chef. If the restaurant you're targeting only has cooks, scratch them off your list.

Respect their time. Just as you are conscious of saving your busy customers time, arrange to see chefs between their busy meal preps. Generally mid afternoon is a good period, after the busy lunch service and before preparation for the dinner crowd. Take a cooler of products for the chef to experiment with.

Education. These days more chefs are aware of and may even be using some grassfed products for their menus. However, don't assume they know everything there is about the health benefits and superiority of grassfed products. Be prepared to offer insight and education, and leave behind flyers or brochures to reinforce your face-to-face discussions. If the target is a white-tablecloth, upscale restaurant, the chef may be more open to using quality ingredients.

In the wings. Similar to an actress before going on stage, rehearse what you plan to say. Pick three key points you want to make. Be clear, concise, and convincing. Let your passion for what you do shine through.

Price objections. Some chefs are solely concerned with using the least-cost products with the highest-priced margins. They most likely will have a purchasing system and company in place where they obtain all of their ingredients on a regular schedule at discount rates. This can be a tough wall to break through, particularly if the chef is a "celebrity" chef in his location and assumes he doesn't have to do anything different

Creating a press kit

Leave a press kit with information chefs can read at their leisure.

The first page should have a quick overview of who you are and what you have to offer. Be sure your contact information is there and easy to find. A photo also helps.

Include your best clips of stories that have been printed about you. In the beginning when you don't have any clips, create your own. Pretend you're a columnist for your local paper and write about grassfed benefits. Give the article a headline and yourself a byline. Set it up in columns to give it the appearance of a newspaper or magazine article. Make sure it's information that reinforces the quality and benefits of your products.

Assemble all of the information in an attractive binder. A clear plastic cover will make your name harder to overlook.

to keep the customers coming.

As a rule of thumb figure double the price of ingredients to cover all overheads built into the end price on the menu.

Joel Huseby of Thundering Hooves Ranch in Walla Walla, Washington, does a cost breakdown for chefs he approaches in the Seattle area. He calls it "Candy Bar Economics" and is something he learned in college. If the price of a candy bar is 50 cents, most likely you will have that much change in your pocket and can buy it. If the cost is 85 cents, again you could purchase it. The price increase was only 35 cents, but the percentage increase was 59%. "From a producer's persective," he explained, "if the cost of creating and selling the candy bar was borne within the original 50 cent price, the price increase is pure profit!"

Joel said you'll burn your bridges with a chef if you try to sell them the cost of a case or the price per pound rather than the cost per portion. Chefs will take the cheaper route with the "natural" vendor. Instead, approach them this way:

"If the price of our ground beef patties is $4.99 and he is calling for a 1/3 pound patty, the portion or per burger price is $1.66. Let's say the competing product sells for half that, $2.49 or $.83 per burger. The difference is $.83 between the lower quality generic product without a story and yours.

"Let's assume," he continued, "that the generic burger currently sells for $7.50. (Sorry, but we don't compete with the franchise folks. This is a great burger!) Now, if we convince him to sell our burger for $9.00 instead of $7.50, the chef more than covered the increased 83 cent cost of ours without losing clientele, because those folks who are our customers and can afford and appreciate our burgers probably have the money in their pockets. For the chef it now becomes more profitable than the generic he was selling previously."

Joel has found that once the chef purchased their grassfed product and put it on the menu, it became the most-ordered menu item and the chef's most profitable one. "Don't just sell your product," he said, "BE your product!"

Remember the example of grazier Ed Martsolf in Morrilton, Arkansas, who does something similar with his grassfed lamb. He compares it to a candy bar, which is about the same size as an ounce of meat and reminds his customers how much healthier the meat is than the candy bar.

Demonstrations. Seeing is believing. Bring a skillet and cook an equal amount of your grassfed ground meat (or other product) and theirs. Then pour off the greasy excess of theirs to show the difference. That alone is proof that your meat offers more value and meat to justify your price. You can do the same with one of their commercially plumped and water-injected chickens compared to yours fresh from the field.

The proof is in the baked goods. Eggs raised on pasture with access to plenty of bugs are one of the easiest things to sell to chefs. All you have to do is crack one open to see orange-rich yolks compared to their pale and watery counterpart. Allan and I were on a cruise when one morning another guest ordered a fried egg for breakfast. It had a rich orange color prompting the other diner to comment, "What's wrong with this egg?" Allan couldn't help himself from saying, "What's right with those eggs? That's the color they're supposed to be."

Chefs who are renown for their baking can't resist using pasture-fresh eggs. They make better, richer cakes thanks to the quality egg ingredients.

Do it yourself. I recommend making the initial contact with the chef yourself. However, you may have hired someone—a family member or non-family employee—to help with your marketing and/or make deliveries to restaurants. If you must send someone as your liaison with the chef, make sure they represent you professionally. They should dress attractively and know your message by heart. If they are not as passionate as you are about your grassfed products they're not the correct person for this position. Once you've established a solid relationship with a chef, someone else can take over routine deliveries.

Delicate approach. David Evans of Marin Sun Farms near Point Reyes, California, had his product picked up by Alice Waters of Chez Panisse thanks to the buzz from the local food scene. He said giving them product at farmers' markets works well. Then follow up within a week to get their reaction. Always get their contact information —phone, email, everything. And stay in touch. "You have to get your product into their mouth AND it must taste good. Once a relationship is ruined, it's hard to get them back." He stays in close touch with chefs to build a relationship and make changes based on their feedback. Because he had far too much to manage, he eventually hired an experienced sales staffer dedicated to creating lasting relationships.

The farm as destination

Radio personality, Studs Terkel, concluded in *Working*, "We have a right to ask of work that it include meaning, recognition, astonishment and life."

Let's add one more characteristic—that it be fun.

Learning something new energizes us, which in turn leads to enjoyment of our work. When you love what you do, your customers will notice.

If you're an outgoing people-person, you may want to look for ways to bring people to your farm. The easiest place to start is to make the on-farm store a destination in itself. Use the store to launch periodic farm tours or special events—Easter egg hunts, Harvest Season hay bale mazes. (Allan would offer train rides.) School groups are always looking for educational outings. Urbanites are seeking authentic, rural experiences and seeing where their food comes from.

Special event. Start small with a one-time event. If you enjoyed it and can justify the time, energy, and results, turn that into an annual, semi-annual or quarterly event. People remember peak experiences, so give them something memorable to talk about it. Word of mouth will enable the event to grow with little to no advertising.

Steve and Dorothy Enger host Fall Family Fun on the Farm with an educational corn maze—one year it might be in the shape of their home state of North Dakota, another year the

USA. Over 5000 people make their way to their remote farm each year.

How about offering stargazing during spring and fall equinoxes?

Explore your heritage. You're promoting healthy, grassfed products. Done well, they should evoke the nostalgia of eating delicious meals your grandparents prepared and/or produced. Was your family German, Swiss, Italian, or Irish? Take your visitors back to immigrant times.

Dairyman, Joe, and his wife, Therese Hayden, in County Wicklow, Ireland, started out offering farm walks and tours to other Irish grass farmers. They built a small visitors and conference center on their farm and decided to entertain foreign guests with a true Irish experience.

They recreated a village from 1938 complete with vintage tractors and cars, a pole with directional signs to "Black Tom's Pub," and "Fair Green," the village center. They built a church facade to hide a drab barn and held a wedding that caused some to ask, "Was that a real wedding?"

Straw-rope making demonstrations, a blacksmith shoeing horses, and a family of traveling tinkers selling their goods were just a few of the living history vignettes created by a troupe of actors, musicians, dancers and crafts people. Visitors played traditional Irish games—welly (rubber boot) throwing and pitching the sheaf—and interacted with a butter maker, bread baker, weaver and more.

Near Ardrahan in County Galway, Ireland, Rathbaun Farm regularly prepares a traditional Irish farmer's lunch with a bread making demonstration for bus loads of tourists. But first tourists watch a sheep dog round up a small flock, see various sheep breeds in the barn, and they can even try their hand at shearing.

Near Dixon, Nebraska, Rose and Larry Mason of Tarbox Hollow Living Prairie Ranch have educated over 6000 visitors a year with covered wagon tours. A small restored building is used to welcome guests who learn it is the same size as a typical one-family residence of the 1860-1880 era in Nebraska. Rose dresses the part in a floor-sweeping skirt as she talks about the hardships pioneers endured in settling Nebraska's prairie and the importance buffalo played in the prairie ecosystem. Exhibits show prairie wildflowers and native grasses, which visitors can search out during the tour.

A tractor pulls a covered wagon to see the pastures and buffalo, which can be fed range cubes. This allows them to see the extremely long, blue tongue and get close up photos of the animals. Back at the homesite Rose spins buffalo fur (they shed their winter coat as large patches of cashmere-like fur). And guests can purchase a roasted buffalo-meat sandwich plate. "Many visitors have told us they came because we weren't a tourist attraction," Rose said, "They came because we are the real thing."

If it's fun for you and your visitors, make the event as rich as cheesecake.

Share your interests. What do you enjoy doing in your spare time? I know, many of you are asking what spare time? If

Taking inventory

Identify all of the farm resources from the ground up—soil, water, plants, animals, people, and don't forget the sky for stargazing.

Wylie Harris suggests answering the Four Ps to create a marketing plan. What's the product, price, place and promotion?

"Does the product match an existing demand? Is its price low enough to attract customers, but high enough to yield a profit? Is the business easily accessible to customers? How will the target audience be reached?" he wrote in "Field Notes," the Kerr Center for Sustainable Agriculture.

you think you don't have time to pursue your favorite leisurely interests, then make that a part of your farm's field day.

Duke and Janet Phillips of Chico Basin Ranch near Colorado Springs, Colorado, love art. For several years they held Artists' Gatherings of exhibitions and sales in the fall. When they ran out of easels for displaying framed paintings, they propped artwork on shelves of hay bales. They asked the artists to interpret their feelings for the ranch in order to communicate that to the outside world through art. In turn, the artists brought together people from different backgrounds to learn about the role of ranching today. The event honored their mission of increasing awareness of land conservation and preservation, showing how the ranch's beauty and the economy can blend well together.

At A Diamond Ranch in Oracle, Arizona, Greg Vinson offers a smorgasbord of activities. River kayaking lasts an hour, just long enough to appreciate the scenery without getting sunburned. The Arizona Trail runs through their ranch. Due to liability issues, they don't offer horseback riding, however they do use small burros. Guests lead the burros and the burros are packed with a picnic lunch. Some visitors enjoy plant identification tours. Others want to explore by Jeep. They renovated the ranch house to accommodate birders. And they teamed up with a balloonist to offer balloon rides over the ranch. This win-win idea provides extra income on top of the balloon ride fee, and the balloon lady can cruise in comfort knowing she's not trespassing.

Charge for the view. In British Columbia, Canada, Lyle Young of Cowickan Bay turned his barn into an Art for Ewe gallery on select weekends, combining fine arts and farming. A music lover, he hosted a Symphony Under the Stars, complete with a grassfed meal. Capitalizing on the view, he also marketed his farm as "The prettiest place to pop the question," and he included a lavish dinner for two.

Urbanites love pastoral settings for picnics, family reunions, birthday or anniversary parties. Rent tables, chairs, tablecloths and dining utensils and set them up where guests can enjoy the view.

Pizza in the pasture. Heather Smith in Wisconsin bakes pizzas in an oven next to her barn May to October. Once a week pizza lovers drive out to buy a pizza, pick blackberries, or socialize with others while the pizzas bake. She uses vegetables from her farm, cheese and pork from neighbors and organic grain from a mill next door. On pizza night she sets up her own farmers' market table at the barn. This gives her an opportunity to network with customers and show them how local foods can be sustainable.

One of her products is raspberry preserves, which she sells at the local co-op. At first her preserves were shelved with everybody else's and didn't stand out. She talked to the manager and explained this was a local, fresh product, "summer in a jar." She convinced the manager to place her preserves near the fresh produce, which caused sales to jump.

The oddball suggestion. Here's an idea from *The Wall Street Journal*—pasture golf. According to the pasturegolf website, golf was originally "a game rural people played simply on the fells and fields where cattle and sheep grazed." These rough, but ready courses open the game to the average Joe—no elitism or high greens fees here. In fact, some holes are so "natural" that there's no space to putt.

Devotees use a different set of rules as they play around snakes, shrubs, telephone poles, livestock, and irrigation obstacles. Some of the unusual hazards are a helicopter pad between the first and second holes at Bear Valley Meadows Golf Course in Seneca, Oregon, (pay your greens fee at the nearby Mini Mart); a resident herd of elk at Salt Creek Country Club in Midwest, Wyoming; and the bang of mining at Pend Oreile Golf & Country Club in Metaline Falls, Washington.

Tom Sawyer working farms. Frank Randall in Alabama said people have offered to pay him five dollars just to stand outside the fence and watch him mustering sheep. Run through your day and all the tasks you normally take for granted. Much of this is new and exciting to families who don't live on a farm.

Some farmers offer bed and breakfast accommodations, plus the opportunity to work for your supper. According to *Gourmet* magazine, Maverick Farms in North Carolina's High Country has 65 acres mainly devoted to vegetable gardens. In their dedication to sustainable agriculture, they use meat from farmer friends, plus butter, milk, ice cream, yoghurt, juice and bread made on the farm or locally.

They host "farm dinners," work days and cooking classes. College interns can earn credit by working on the farm, and overnight guests can reduce their lodging bill by the amount of work they do on the farm. None of the paying guests are required to work, but the hope is that they will pitch in and lend a hand. Because the owners speak to potential guests ahead of time, most who pay to work share their vision and interest in sustainability. The advance phone calls helps avoid high-end guests with unreal expectations.

Blackberry Farm in Walland, Tennessee, has become so famous as a culinary center, it belongs to the prestigious Relais & Châteaux listing of luxury inns. Guests here aren't asked to work, but are invited to watch the sheep being milked or the cheese being made. Guests pay nearly $1000 (and upward) a night with a two-night minimum. Something to strive for.

Destination event. Renaissance Fairs, the Society for Creative Anachronism, and Civil War reenactors among others need plenty of space to recreate their special worlds from the past. This probably won't work if you have cannons booming near a pasture full of livestock. However, if your farm is extensive in acreage and can keep the pretenders separate from the grazers, the event can be a crowd pleaser and income booster.

Maybe you won't hold the event on your farm, and you will only feed the troops with your grassfed products. Remember, demonstrations alone won't sell anything unless you can get people involved. Give people a "taste" of the past.

Is there a Medieval Banquet near you? Since forks had

not yet been invented during the medieval era only knives are used in today's recreations. Whole chickens are standard fare for roasting. And there are less utensils to wash later.

Summer camp. Turn your farm into a summer camp with week-long educational classes. Include plenty of hands-on opportunities to forage for or simply identify edible wild plants. People like a schedule to give structure to their day, however, keep it simple with an easy 9 to 2 program to allow for leisure time.

Pay Dirt Farm School is a non-profit educational program of Mary Jane's Farm near Moscow, Idaho. Apprentices learn how to plant, grow, harvest, preserve and store healthy foods. They're also challenged by typical farm tasks—repairing broken equipment, mending clothes, setting up and moving fences. A sampling of programs is Building a Backyard Greenhouse and Grow Your Own Herbal Medicine.

The Farm Institute at Katama Farm on Martha's Vineyard has daily and week-long summer activities for children of varying ages: Wee Farmers ages 2-4; Sprouts ages 4-5; Growers and Harvesters ages 6-13; Homesteaders ages 9-15; Future Farmhands ages 11-15; and Farmer for a Day ages 6-15 with their parents. Each week is themed as a COW—Concept of the Week—such as land preservation or humane treatment of animals. A kiosk holds cards for Livestock Quest—a self-guided farm tour, Community Chores, Hayrides, and a Farm Stand selling hats, T-shirts and produce and meat.

Take home lesson. Remember, what you take for granted and think of as ordinary can fascinate city folks. Show what you know. Always have something people can take home with them—blueberries, honey, brochures, flowers, recipes, grassfed products. This keeps them thinking about you and gives them a reason to return to buy. Put your name on everything so they will come back to buy more.

Farm tours, how tos

Farm tours are an excellent opportunity to educate consumers. Create posters and signs everywhere you take the tour to educate them on environmental issues, healthy animals, beneficial grassfed products. Think of it as on-farm advertising and community building. Think of the farm as a blank canvas to paint all of these relevant ideas.

Do the numbers. Unless you have a large family or lots of friends to help, you won't want to offer farm tours or open field days too often. They're a lot of work and take months of planning. The Salatin family's Field Day in Swoope, Virginia, has grown to 1300 visitors and is held every three years. Their audience is comprised of 1) other farmers who want to see for themselves the famous Polyface Farm and learn first hand how the Salatins operate and manage their multi-species enterprises; and 2) customers who enjoy a day in the country.

When they conducted farm tours for school children (school districts have since nixed off-site programs), dairyman, David and his wife, Leianne Wright in Alexandria, Alabama, enlisted family members. They staggered tour times between nine and noon so that they could tour 300 children in one morning. Aside from the tour fee per child, product sales soared 30% on tour days.

Infrastructure. Having parking areas and roads adds to repeat business. In searching for designated places to park look for areas that will receive the least amount of damage to pastures and property. Be sure there's a bathroom (portable for large groups) and shelter in case of a sudden change in the weather. Don't forget liability insurance. Don't forget to set out several garbage cans.

Schedule. Be sure your facilities and pastures look their best. Work around your work. Schedule tours on the slowest time of the week or year. On the other hand, if how you do something is part of a field day tour, you may want to leave the work until you have an audience, and potential volunteer farm hands.

The enforcer. Set the tone for control of the group up front. Enlist the aid of family members or friends as well as teachers and chaperones who arrive with school children. Keep everyone together to help solve some of the liability issues.

Knee high. When your tour customers are children, get

Practice makes perfect

Before inviting guests onto the farm, take the tour yourself. Be sure you've allowed enough time, but not too much. It's better to pick a theme to focus on for the tour rather than try to cover everything. That way you can change the theme to encourage return visits.

Look at the farm as the first-time visitor will see it, then fix any problem area.

Summarize the key points you want your visitors to know. Then allow plenty of time for questions and answers.

Offering samples sparks their interest in purchasing.

Escort your visitors back to their cars or buses. Invite them to return and tell their friends.

down on their level when you talk to them. This is particularly important if you're trying to teach them something.

Show don't tell. Telling them what you do is not nearly as memorable as watching you and your animals in action. Have your sheep dog round up a group of lambs. Let your guests watch sheep being sheared. Save the cattle's paddock shift until you have an audience.

Pricing. If you charge an entry fee, make it simple to eliminate the need to make change. $1 or $5 is better than $3. Better yet, don't charge an entry fee. Let people buy tickets with a dollar value each. Once they're on the farm, one ticket allows access to the petting zoo. Five tickets buys a hamburger. Another ticket purchases a hay ride.

Enthusiasm. Make your visitors feel as if you are glad they're there. Treat them like members of your extended family. And don't send them home empty handed.

The farm stay

Several years ago an innkeeper told me, "I never have been able to travel, but by having a bed and breakfast, the world comes to me."

Operating a bed and breakfast is a business in itself with a whole set of regulations, headaches, and yes, joys. It can also be an adjunct to the farm by bringing in customers for your produce and grassfed products.

Being the innkeeper means your farm gate can be open —or closed—whenever it suits you. Maybe you want to have guests a few times a year when other farm duties are light. Or maybe you love entertaining so much you keep the welcome sign out year-round.

Let's look at slices of this farm pie:

Back to nature. Urbanites love being in the country, as long as the environment is a safe one. Your farm can allow them to interact with nature without worrying about trespassing on posted property, walking into some hunter's scope site, or getting lost in the woods. They want to breathe in the clean air, see stars in the sky at night, get away from the city's noise. And they want their children to experience this too. All those resources cost nothing to share, yet simple stargazing events, guided hikes, wildflower identification can add off-season revenue.

Seasonal getaways. If you plan to welcome overnight

guests, accommodations don't have to be fancy. A friend raves about a favorite cabin in Gatlinburg, Tennessee, that backs up to a stream where she can listen to the burbling water and frog songs at night. "It's not any bigger than a two-car garage, but it has a fireplace," she said, adding that she has been going there for nearly 20 years.

Becky and Paul Kelley who offer hunting leases on their Sylvanus Farm in Nebraska converted an old railroad car into a guest room.

One western grass farmer built a lodge where meals are served and showers are located. Clustered around this hub are covered wagons for sleeping accommodations.

Should you offer Bed & Breakfast lodging?

If you are planning on adding a B&B to your farm business, here are some steps to take:

1. Check local zoning and ordinances.
2. Read a book on innkeeping or take an online course.
3. Ask yourself if you have the correct personality for the role. Do you enjoy entertaining? Do you get up early? Do you like preparing meals?
4. Evaluate how a B&B will affect other members of your family. Will the role of innkeeper interfere with family schedules?
5. Talk to B&B owners.
6. Sleep around...at other B&Bs to gather ideas on what's done well and what is lacking. If you're in the B&B business, this "research" can be tax deductible. Check with your accountant.
7. Have you factored in time out for yourself and your family?

Indian style tepees infuse a spirit of adventure and give city slickers plenty to talk about around the water cooler at the office. Even deluxe African safari companies use tents for guestrooms. Let your farm be an alternative to state and national parks by providing a designated camping area for primitive and RV camping.

Horse enthusiasts know that people trailer their horses around the country to exercise them in different locales, particularly prior to horse shows. Why not offer a B&B&B? That's Bed and Breakfast for owners and Barn stabling for their horse. They'll be grateful for a place to ride without worrying about trespassing on private land or dodging hunters.

Buy binoculars. Urbanites love to escape to natural settings. For many the more remote your farm location the better. They're thrilled to catch a glimpse of deer in a stand of trees, turtles plopping into a pond, or migrating birds.

The innkeeper of Fort Lewis Lodge, a 3200 acre farm in the Allegheny Mountains near Millboro Springs, West Virginia, escorted city slickers on guided "hunts" with cameras. They built wide-platform tree stands to view wildlife in wide-open pastures. However, the highly anticipated event of each day was the meal the owners prepared from their own farm and served in a converted grist mill.

Remote Rancho Mababi near Aqua Prieta, Sonora, Mexico, lies on the southern flyway for migratory birds—a birder's heaven. Although Alice and Roberto Valenzuela knew nothing about birds, they invited a group of birders to visit the ranch in exchange for an inventory of sightings. Then they subscribed to *Bird Watchers' Digest* to learn the interests of how to reach these customers, who typically have high income levels.

Is your ranch located where Native Americans settled? Are there petroglyphs or pottery shards? Guide your visitors to these hidden treasures.

Children aren't the only ones who enjoy learning something new, especially if there are no tests. Select the prettiest time of year on your farm and invite a photographer friend to conduct a photo workshop.

Keep a guestbook and you'll soon discover people from around the world knocking at your door.

Is there a festival in your future?

In Europe during the Middle Ages, Guild Fairs resembled today's family reunions. People traveled far distances to the fair town. They bartered and sold goods they worked all year to produce. Capitalizing on the crowds, jugglers, stilt-walkers, and other entertainers flocked to the site. It was more than commerce. It was a celebration.

At Brazos de Dios community located west of Elm Mott, roughly midway between Dallas and Austin, Texas, the Homestead Craft Fair celebrates the end of the harvest season and the culmination of the homeschooling year. From its first event more than 30 years ago when it attracted 1500 people, it has grown into a three-day heritage festival of music, hands on children's activities, demonstrations and displays of cheesemaking, shearing and spinning, cow and goat milking, woodworking, pottery, leatherwork, and basketry, and that's just a sampling of everything that can be seen and done.

"People come at 10 am, eat three meals, and stay for cider around the evening campfire," said Butch Tindell. Since many visitors do their Christmas shopping at the Fair, sales spike and continue until Christmas, accounting for 25% of the community's annual income.

Along with the heritage crafts, attenders experience grassfed meat prepared as American, Italian, Mexican and Mediterranean meals at the 75-seat deli. At peak times there's an hour or more wait to buy food. Fifteen thousand people (an average of 5000 each day), purchased 23,000 grassfed meals in 2008. Many more bought grassfed products at the community's store to take home for their personal consumption.

Is there a festival in your future?

Here are some of Butch's suggestions for creating your own festival:

Choose a theme. Have a clearly defined focus and be sure it's family oriented. For example, with Brazos de Dios as the centerpiece, Allan spear-headed a Heritage Food Festival for our home town of Hattiesburg, Mississippi. Our festival is designed to celebrate the pre-World War II food heritage of South Mississippi. Under that general theme, each year the festival can have a different food focus, beginning with sweet potatoes for the first event in 2010.

Changing the focus each year guarantees there will always be something new to draw people back again. Our church has an Irish-Italian Fest that begins with a Celtic bag-piper leading children and their parents into the fair area. Each year something new is added, which has turned this into a community event.

Early, thorough planning. Begin planning at least a year in advance. Now is the time to determine the parameters. Do you have the space or location to be as big as you'd like it to be? Will it be on your farm or elsewhere? Check local ordinances about rules for serving food. In many instances regulations are waived for a one-day event.

Start small. Like a thousand-year-old live oak that sprouted from a marble-size acorn, build on something you're already doing to get people to your farm or event. For years Scott Blackman from Wolf Creek, Montana, has drawn families to his 20-acre corn maze. He started with a professional maze designer and a different maze and theme each fall. Now he is planning to create the maze himself and add more reasons for families to come more often, stay longer, and purchase more. Children's activities always attract more families.

Inventory resources. Identify people, organizations or groups who can be part of your planning team. The Convention & Visitor Bureau, City of Hattiesburg, Historic Downtown Association, and the University of Southern Mississippi jumped onboard for the Hattiesburg festival. They were quick to key off of the 100th anniversary of the university and the town's 100-year-old restored train depot. For our festival, liability issues fell under the umbrella of the Convention & Visitor Bureau.

For our church festival we hire Youth Challenge at-risk kids to help with the set-up, decorating, tear-down, and trash collection details. Giving them specific areas of responsibility enhances their trust and self-esteem.

Get organized with everyone on your team. Some work can be done by email or telephone communication, but you will also need face-to-face planning meetings.

Funding. Are there like-minded grass farmers who can help share the costs and the benefits? Solicit business sponsors in your area. Check into a SARE grant. Look for complementary musicians or entertainers whose talents would enhance the experience. A friend of ours has an oyster bar where he invites musicians to play for a very small fee. They know it's exposure for them and an opportunity to get other gigs from someone in the audience.

Artists and crafters will often exhibit their work in hope of a sale. If you charge for booth space, keep the fee low initially. A word of caution: If you invite artists or crafters as part of your event, make sure their work is juried to keep the quality high. Be sure their expertise fits your theme.

Location. On the farm, tents add a festive atmosphere and give you a tented-in backup rain plan. However, the farm may not be the best place for your event. Search for historic public buildings, parks, a building not in use year-round, or even a church parking lot.

Our church started the Irish-Italian Fest to celebrate the ethnicity of our parish. A neighboring church has an annual

Oktoberfest in their parking lot, which is so small many people come simply for take out meals. (Notice both are food events.)

Advantages of farm festivals

Taste and sell. This is an opportunity for people to sample and buy grassfed products. Free samples equal a sales boom, especially for people who are new to grassfed and are reluctant to purchase without first knowing what they'll be getting. People are more likely to pay more for food at a special event.

Do the numbers. Any event, regardless of size, will sell a lot of product. That means you must prepare months ahead to have enough product to match anticipated demand.

Community building. Brazos de Dios has the backing and workforce of the entire community, yet it relies on the cooperation of neighboring farmers for its success. Wheat farmers augment their own supply to make brick-oven fired breads. A nearby dairy supplies additional milk for cheesemaking. The fair has spurred spin off business startups in the community.

Sustainable ag. Such an event raises awareness of sustainable agriculture. It shows city folks the kind of healthy families and lifestyle that come from clean water, lush pastures, pretty landscapes, and happy animals.

Education and awareness. Throughout the day you can include educational "food bites" on the differences between grain-fed and grassfed products, how to prepare grassfed products at home, or what consumers can do to make their environments "greener."

Infrastructure. Restrooms are a must. In addition to Port-a-Potties, you can rent trailered restrooms similar to dressing rooms used by movie stars on location. Depending on your theme and focus, maybe you want to build outhouse facades over the Port-a-Potties. Don't forget to cut out a half moon for the door. Hang a catalog on a chain to add a touch of whimsy to that old-fashioned theme.

Rent a sound system for your music, but also for those announcements of a child who has gotten separated from their parents. For our church festival we built a signpost in a bucket of concrete. Placards pointing the direction of various activities are attached with velcro so that they can be removed and reapplied every year.

Security and parking. Golf carts and drivers shuttle fair-goers from off-site parking for our church event since the fest takes up the entire parking area. If your event is on the farm, people will park in your pastures. Make sure they use your least productive areas because the ground will be packed down from the traffic. Make arrangements with neighbors to lease parking space. Look for offsite areas where shuttle buses, golf carts, or hay wagons can transport people back and forth.

Contact your local sheriff, police force, highway patrol. They enjoy being involved and can help with the traffic crunch. For our church's Irish Italian Fest we pay the security staff and workers in free tickets that can be used for food or activities. Be aware that serving alcohol may draw more adults with cash in their pockets, but it also has the potential to compete with the family atmosphere and may result in disruptive behavior.

Education. Capitalize on your captive audience by giving them opportunities to learn more about you, your farm, your family, and your grassfed products. Repeat the same talk several times a day, and post a schedule. Don't just offer games or blow-up bouncy trampolines for children. Give children hands-on options such as building a bluebird bird house or making a stamped leather key chain, dipping candles or knitting. Think education.

Promotion. Send out press releases to area and regional media outlets, which often result in free publicity. Stress that your event is a "locovore" with food sourced from the community. Then stand back so you won't be trampled when the crowd arrives.

Professionalism

Database marketing

Database marketing is knowing who your customers are by their names and addresses, however, to be most effective you need to go beyond names or lists.

You need to know what your customers buy, how often, how much they spend. Database marketing identifies the 20 percent of your customers who account for 80 percent of the sales. Keeping these records on computer allows for frequent updates.

Here's the information you need to capture:

1. Name of customer.
2. Address—mailing, email, shipping, and/or buyer club delivery site.
3. Telephone number—home and cell.
4. Time zone (if you are shipping orders).
5. Source of inquiry (Flyer, word-of-mouth, public relations).
6. Name of customer if they are a referral. Be sure to send a thank you or give your best customers a little something extra in appreciation.
7. History of purchases:
    - a.) by date
    - b.) by item
    - c.) by dollar amount
    - d.) by cumulative sales

8. Testimonials and permission to use them in your marketing materials.

9. Notes of family members or other pertinent information such as birthdays and anniversaries. Knowing these details and using them in conversation adds a richness to your relationship with your customers.

Follow up. If you want repeat sales, you have to have a plan to follow up after the first purchase. Follow up reminders build trust and reliability and initiate a relationship between the seller (you) and the prospect. After your prospect has left, your follow up post card or email will jog her memory to start thinking about your products and anticipate buying from you...next time.

Your customer has just thrown the ball to you. Don't drop it by ignoring the opportunity to follow up. Jay Conrad Levinson of the *Guerrilla Marketing* books says that apathy after the sale can cost you 80% of business.

Timing. Frequency in sending missives can make your product stick in her mind. However, there's a fine line between

How can you profit from follow up?

In *Guerrilla Marketing Weapons*, Jay Conrad Levinson gave this example of how profits can add up from the simple act of follow up:

A $200 profit becomes $600 annually when that customer buys two more times.

Four referrals from a satisfied customer (each buying three times a year) totals $2400 annually. Add that to the initial customer's sales to get $3000.

Multiply that by the number of years you keep those customers and you can see the dollar value of follow up.

gentle reminders and pesky messages. Make sure you have something worth saying.

Stand out. If you plan to send cards or email, pick a non traditional time to allow your message to rise above the clutter. Instead of Christmas or Easter cards, send out Ground Hog Day cards or Chinese New Year notes. Send a thank you note on the anniversary of when your customer first became your customer.

Promotions. If your budget allows, include promotional products emblazoned with your farm's name—coffee mugs, blank recipe index cards, post-it notes in animal or barn shapes. Email a sample newsletter. Create a generic newsletter so that you won't have to update it too frequently. Once the prospect becomes a valued customer you can then send regular newsletters about your products.

Keep it personal. Drop a handwritten postcard in the mail thanking them for their purchase. If this was a referral, write a thank you note to the customer who sent them to you. Create inexpensive postcards with valuable information or tips on one side, the address and room for a short handwritten note on the other. Send a recipe and invite your customer to share one with you. Chico's, my favorite clothing store, always includes a handwritten note whenever they ship something to me. Sometimes it just says, "I enjoyed helping you with your selections." It makes we want to seek out and shop with the same sales clerk next time.

Make a call. If you've established a stronger friendship with the customer, pick up the phone and ask how the dinner party turned out. Call before they start fixing that rack of lamb to ask if they have any questions. When you learn something about them from your conversation, call them to let them know you're thinking of them. For example, my neighbor knows I collect hippos. She will buy one for me when she sees them, or if it's a pricey one, call to tell me where she saw one. Maybe you and your customer were casually talking about winter planting. Let her know you saw cyclamens or pansies on sale at the nursery so that she can shop early for the greatest color selection.

Housekeeping. Once you have a list, assign someone to keep it updated. Delete customers who have moved. Correct those who have changed addresses or other contact information. Periodically purge those customers who haven't purchased anything in more than a year. Don't lose those names, simply created a second database with these lost customers so that you can target a special "come back" marketing message to them.

The more you do to show your customers you care, the longer they'll continue shopping with you.

Giving thanks

According to a newsletter called "The Motivator," there are eight circumstances when you should thank you customers.

1. When a customer does business with you, every time!

2. When he or she compliments your business, farm or ranch.

3. When your customer offers comments or suggestions. Acknowledging feedback shows you're listening and that you value their opinion.

4. When he or she tries your new product or service.

5. When they recommend a friend to you. Timeliness adds to your sincerity.

6. When your customer is patient, and even when they aren't. Thanking them shows you value their time.

7. Some customers are always prepared. When they make your life easier and help you serve them better, thank them for it.

8. When they complain. Expressing their unhappiness is a sign that they're giving you a second chance.

How can you say, "thank you"? Let's count some ways:

Bonus. When someone refers a new customer to you, give them something extra: a free dozen eggs, one more pound of ground meat, a shopping bag (with your logo and phone number), or flowers. Don't discount your products as a way to give thanks. However, a discount or perhaps waiving a shipping fee, may be your policy for the organizer of a buying club.

Snapshot. Take their photo (holding their purchases) and post it in an upright plastic display frame. Title it Hero of the Month or Our New Friend, whatever you can think of to make them feel special and pique others' curiosity. Perhaps you draw names from business cards you've collected.

Notes. Create postcards or one-fold cards with a pretty scene from your farm. If you choose to have a printed message on the card, be sure to handwrite a personal note as well. Handwrite the envelope too.

Phone. If someone drops out of your regular buying circle or a buying club, phone them to thank them for their patronage and to find out why they're leaving. This could reveal valuable information of something that's not working, which you were unaware of happening.

Heads up. Give them first notice of a popular product in short supply, and be sure you let them know they're on the priority list because they've been such a good customer.

Origin of the word sincere

Sine, "without," and *cere,* "wax," are two Latin phrases used by stone craftsmen during the Republic of Rome. To cover imperfections in the marble they worked they would use wax, similar to rubbing wax into defects or scratches in antique wood furniture.

Some suggest that unscrupulous brick layers would use wax rather than the more expensive cement between bricks. When it melted, the bricks would shift and the structure collapse.

Claiming that something was *sine cere,* without wax, became a guarantee of the real thing.

Appearance

Growing up, I attended a school where uniforms were mandatory. Every day I donned a pleated navy skirt, white blouse, white socks and saddle oxfords. In winter a navy sweater completed the ensemble.

I hated it.

However, there was an up side to wearing uniforms. The dilemma of what-to-wear-today was solved, allowing me a few extra minutes of sleep before getting dressed. My school clothes never broke my parents' budget in a race to keep up with wealthier Joneses. Whenever someone got in trouble, sometimes they could blend in with everyone else, especially when there was a substitute teacher who thought we all looked like clones. And, it identified all of us girls with our school, which meant we stood out from everyone else in town.

When it comes to selling grassfed products, should you wear a uniform? If so, what should it look like?

Several years ago I was in Atlanta on a site inspection for a hotel for one of our conferences. I'd pretty much made up my mind about the hotel to book. It had all the amenities, space and rates to meet our needs. It was late in the day and I had one more stop to make at a hotel then managed by the Harvey Hotels chain.

From correspondence and phone calls prior to my trip I knew the meeting space in this final hotel was smaller than what we wanted. However, I was impressed with the corporation's can-do attitude and empowerment of their employees: if you asked the janitor for a slide projector he wouldn't say, "That's not my job," he'd say, "Yes, ma'am, I'll get it for you," and he would find the correct person to handle my request. So, I had agreed to take a quick look.

When I drove up to the hotel at the appointed time I was floored to find all of the managers on duty lined up outside to greet me. They had rolled out a *green* VIP carpet from the door into the lobby. This alone showed they had done their research about the *Stockman Grass Farmer*. Inside I was greeted by one of the managers dressed up as Grassy the Cow. Their sense of humor, thoughtful research, and outrageous, out-of-the-box effort won our business. I didn't even have to look at the meeting space to know these were our kind of people. We held a very successful meeting in a somewhat tight space, and no one complained.

I'm not suggesting you dress up in sheep's clothing to sell lamb (well, maybe at Halloween, and only for fun). What this anecdote says is that you are what you wear. Your appearance makes an unspoken statement about you. It can imply social or economic status, a sense of power or control, like ability and credibility. It can even be a subtle tool to win over a customer.

Lookalikes. Brent Walter, sales and marketing director for Fox Fire Farms in Ignacio, Colorado, suggests you should dress like your customers. "Your potential customers will size you up in 30 seconds or less," he said. The first thing they see is what they will concentrate on. If you look as if your clothes are fresh-from-the-laundry clean, they'll assume the food you are selling is fresh and healthy (even if it's frozen).

Generally the Fox Fire team wears jeans or khakis like their casually-dressed Durango customers. Other Fox Fire markets are frequented by customers in cowboy boots and hats, so when they go to those markets they dress in western wear.

Urban cowgirls. In cosmopolitan markets (and some

western ones), urban women view cowboys as unhealthy people (thanks Marlboro) who abuse animals (thanks rodeo). Jo Robinson, originator of eatwild.com says to "think tweeds and dry stone walls." She also suggests that if you wear jeans they should be faded in the knees rather than all over like the stone-washed versions. "All of your clothes should be spotless."

Uniforms. Author Faith Popcorn said women are suckers for uniforms. (She obviously didn't attend my school.) That can be as simple as a black t-shirt and black jeans as long as everyone in the booth is dressed the same way. The

Think of your business like a stage

In their book *The Experience Economy*, B. Joseph Pine II and James H. Gilmore suggest treating work as theater.

Different "costumes" reflect different roles. They use the example of airlines where the check-in staff wears different uniforms from the pilots, and the guy with the knee pads is clearly identified as a baggage handler.

Costumes can be used to convey a message. Southwest Airlines' polo shirts and sneakers shout "We're perky! And ready to jump up and down with you (in four connecting flights) to California."

If you decide to have everyone wear uniforms, be an original and create your own standout accessory. At the restaurant chain T.G.I. Friday, the wait staff, who wear red and white striped shirts, is encouraged to wear a hat with slogans and buttons of their choice.

The Experience Economy, Work Is Theatre & Every Business a Stage by B. Joseph Pine II and James H. Gilmore.

"uniform" suggests professionalism and symbolizes authority. A quick glance your way will signal to a customer that this person knows their product and is comfortable selling it. A couple in our town have a small business run by the two of them. They have a variety of different T-shirts in various colors. Each day they both wear the same kind of shirt for uniformity.

By the way, wearing glasses implies intelligence. However sunglasses suggest aloofness. Take the sunglasses off to make eye contact when a customer approaches.

Simplicity. Designer clothes are unnecessary, in fact will come across as being pompous. Khakis and polo shirts present a crisp, confident image. Simple t-shirts or pullovers can be embroidered with your farm's emblem rather than having a Tommy Hilfiger logo on everything.

Show your colors. Is there a particular color identified with your farm or logo? Green is the obvious one, but let's face it, some of us don't look good masquerading as leprechauns. We suggested green for our own staff years ago and were much more pleased with white tees and green lettering.

Some of our staff didn't like wearing t-shirts, so we compromised with a basic look—khakis and white. Some liked skirts and others slacks. We kept our individuality within the same color scheme.

When I worked at one of our Purple Cow conferences, I naturally wore purple outfits. It was neat, reinforced our theme, and it's also one of my favorite colors.

Most women like earth-colored—black, khaki, brown, gray—uniforms best. White works for summer.

Accessories. As for your shoes, just be aware that most people spend more time looking down than up. And jewelry can be distracting. Minimize. A tasteful watch and maybe some earrings - if you normally wear them—is enough.

Well dressed. Don't save your "uniform" just for farmers' markets. Wear them for pictures to go on your website, on market days at the farm, or when journalists want to do an interview.

Clothing designer Ralph Lauren put his female models in man-tailored jackets and pants to give his urban customers a sense of what we already have—a pleasant life in the country. He has been quoted as saying he doesn't design fashions, but rather lifestyle. Similarly, the cosmetics industry isn't selling makeup. They're selling hope.

Putting some thought and coordination into your marketing clothes is not vanity, it's a sign of validity. Planning how you're going to look tells your customer that you care about the quality of her family's food as well.

Pretend to be your customer

Have you ever taken the time to look at your farm business from the outside—the way your customer sees you? They're the ones who are quick to see things you are doing right, and areas that need improvement.

Reverse roles and pretend to be your customer. Pretend you're conducting an emergency drill to evaluate the response and areas needing improvement. Just don't tell your personnel what you're planning. You want to evaluate a typical day, not a special occasion when they might be expecting "company."

Initial contact. Call, write, or show up where your products are marketed. You may have to enlist a trusted friend to be your "industry spy." After all, if your teenager is the one answering and taking phone orders, they could easily recognize your voice. Unless you have a long camera lens like the stake-out detectives on TV, you will have to send someone else to your farmers' market booth or farm store. Make a checklist of things you or your friend should look for.

Phone call. How many rings did it take before they answered? Did you get a recording? Did they identify the farm and their name? Was the person on the phone courteous? Did they sound upbeat? Could they clearly be understood? Did they ask the right questions or supply the correct answers? Did they forget to mention something important? Did they get the correct spelling of the phone caller's name and contact information? Did they follow up with something sent by email or snail mail? Remember, the first voice the prospect hears becomes the "voice" for your farm business, that's why the phone person should be pleasant and properly trained for the position.

Were you put on hold? Don't make your customers listen to dead air or elevator music. Invest in or create your own marketing messages to play whenever someone is placed on hold. Provide grassfed facts, seasonal product offerings, your multiple marketing locations, hours of operation, farm news. Southwest Airlines used to play funny skits that not only kept customers waiting for the human voice to come online, but kept people calling back just to hear the recording.

Write down key points of phone etiquette and post it so that whoever answers the phone in a pinch can easily maintain the professionalism you want to convey.

Write. Write for more information. Again, having an unrecognizable address can help with your sleuthing. What did you receive as a reply to your request? How long did it take to receive that information? Did it adequately address your request?

Was there a follow up phone call to be sure the materials arrived, giving the prospect a chance to have questions answered? Small business consultant, Harold Wright discovered that a followup phone call after mailing a product brochure or catalog increased sales by ten percent. Even if, like me, you're reluctant to make phone calls, a followup call is easy. Unlike a random "cold call" that catches people off guard, this

How to check it out

1. Call.
2. Write.
3. Go by.
4. Analyze the results and implement needed changes.

can be a warm and fuzzy call because the prospect initiated the request for (written) information. This type of courtesy call shows them you're interested in helping them receive all the benefits your product represents.

Drop by. Send your spy into the marketing venue. Is the area clean and neat? Do the personnel look smart and professional? How quick is their response to acknowledging a potential customer? Do they look happy to be there?

Evaluate all the responses and take appropriate, corrective action.

The black box. Harold Wright suggests testing your marketing message by trying to sell a magical black box. You can't see what's inside. You can only tell your potential customer what it will do for them. Remember, concentrating on consumer benefits is the cardinal rule of successful marketing.

Time is ticking. Wright found that two months is the time limit a person could remember the name of a recommended service or product. If a friend recommends a plumber, you will remember that plumber's name for about two months. This is why it's crucial not to delay communicating with your prospects as soon as possible. Lack of name retention makes an on-going marketing campaign essential for sales growth. Even your current customers need to hear from you every two months to keep you in the forefront of their minds. This can be done by email newsletters, or showing up at farmers' markets where they expect to see you.

Grow into your business. Wright said that a marketing plan that can produce results on a small scale will produce virtually the same statistical results when rolled out on a larger scale. That's why keeping records on inquiries and sales' conversations are so crucial to becoming your customer's best friend. Are you or someone in your business doing this?

Displays. Check out your booth from your customer's point of view. How close do you have to get before the banner can be read? What's the first thing you notice? Is anything hidden from view? Do the colors you've chosen for signs and background items evoke a sense of harmony or a mishmash of confusion? What sets it apart from everyone else's space?

What's in the background? Anything that's distracting —coolers, crates, empty boxes—should be covered, otherwise it will give the impression that you're disorganized. Use a checked pattern or disposable tablecloth that matches your key color and cover those necessary but jumbled items. Another suggestion is to position a screen behind you and the extras. Get your children to paint a farm scene on the screen.

One year at a *Stockman Grass Farmer* conference, we were selling books. The table was neat with attractively displayed books, however, the boxes for restocking were beneath the table, and our booth area—the only thing available—was backed by a mirror.

In case you're not close enough to one at the farmers' market (and don't have time to constantly abandon your post to keep things neat), bring your own trash can. And use it.

Second look. I love whimsy and surprise and have tried to incorporate a sense of fun into my home decorating. Create a sense of surprise at your booth—something that will make people stop and do a double take. For example, I have a whimsical, life size, wooden basset hound that is a toilet paper holder. I once used it in a display with a roll of tear-off messages. I've also used my basset hound butler dressed in a suit to hold items on a tray during a dinner party. Be creative and use something that fits your personality as a "place holder" for important information, or to call attention to something new.

Selling by coconut telegraph

The cheapest and most effective advertising is word of mouth. However, this only works with highly differentiated products and services.

It's your job to constantly be looking for ways to get new customers by having your existing customers do the work for you.

"What women buy, women 'sell' when they're pleased with products and services. They will talk about them to others," wrote Marti Barletta in *Marketing to Women.* "Not only is it free, it's more credible, effective and persuasive."

Men vs women. Barletta noted that when you make a sale to a male prospect, that's all you get—a single customer. By contrast, converting a woman gives you her own purchases and a string of lifetime referrals. Women's spending power is more powerful than Japan's entire economy.

Gift certificates. When your customer—male or female—brings you a new customer, reward them. At the *Stockman Grass Farmer* those who refer a new subscriber get two extra months added to their subscription. Names alone don't count, only those referrals who actually follow through with a paid subscription.

Business cards. Anyone can create a business card with a computer and some card stock. However, this takes time and can lead to parsimonious distribution. Having cards printed inexpensively in bulk at an office supply company means you can hand them out to anyone anywhere. Like Pigpen in the Peanuts comic strip, leave a trail of cards wherever you go. Even if they don't result in a sale, they will help people remember you.

When you create business cards, make them work for you. Here are some ways to power up that card from Jay Conrad Levinson and his "Guerilla Marketing Newsletter." Think of them as mini-brochures. Look for unusual ways and places to distribute them—health clubs, beauty salons, day spas, cooking clubs, church bulletin boards, places where people get bored waiting in line. When you give someone your business card, add a second one to pass-along to a friend. Enclose your card with your invoices. Collect a card when you give a card. Ask your customers to share your cards with the right people. List all of your contact information, even if it means having both sides printed. Design a card that opens up and lists all your products benefits. Use your farm logo so that it ties in with your other marketing materials.

Turn customers into converts. Missouri marketer, David Schafer, differentiates between customers and converts. Customers buy your products because they like the taste and want to know where their food comes from. They might also be buying 100% grassfed products because they are chemically sensitive.

Why buy from you?

"If you don't expand the reasons why people should buy your stuff, all you are doing is 'renting volume.' You need to own your volume. New customers are expensive. Existing customers just need confirmation of their behavior and new reasons every day to buy your brand."

Sergio Zyman *The End of Marketing as We Know It*

Converts buy into the whole grassfed movement. They feel as if their purchases are helping to change the world for the better. They become emotionally invested in your products and want to share the experience with like minded friends.

Whenever they find a customer with "convert" potential, they give them a copy of Jo Robinson's book *Why Grassfed Is Best* or Joel Salatin's consumer-directed *Holy Cows & Hog Heaven*, which David regards as a guerrilla marketing assault weapon.

David admits only about 10 - 15% of his customers are converts "armed with the knowledge and conviction to influence others about grassfed." His only tools are free or low cost books backed by quality products.

Public relations

Starting out and trying to keep expenses low, you shouldn't invest mega bucks in advertising. In fact, you don't need to when public relations can serve your interests free of charge. As Mart Martin, former marketer for Coca-Cola™ said, "Public relations comes first. (Paid) advertising should be an add on."

Speaking on a panel at the University of Southern Mississippi, Martin gave as an example the launch of Vanilla Coke. He had discovered the Vanilla Bean Café in Pomfret, Connecticut, (population 3800) and blitzed the business and general media with press releases about the *first* place Vanilla Coke would be tasted. The café, located in a restored barn, only seated 16 when it opened in 1989. It was mobbed by media, who spread the word about the new twist on traditional Coke. Later, advertising helped sustain the public interest in this new product. The Café now seats 90 inside and 40 outside and claims to be "The Official Launch Site of Vanilla Coke."

There's no need to hire a PR firm when you can do it yourself.

Names. Spend some time making phone calls to learn the names of the media people you want to contact. Everything you send out will have more impact if it is addressed to a person rather than just the company. You might discover that joining a business breakfast association will bring you into contact with these media people, which will give you an edge up when they recognize your name on whatever you send or contact them about. Face-to-face trumps name-to-name. Do your homework and get those names before you need to contact them.

Hungry media. The TV, radio and newspaper media is hungry for material. Having tight budgets themselves, they are more likely to welcome creative input and programming, particularly when you can "sell" them on a ready made package that won't cost them anything. Don't expect to be paid in dollars, however, the exposure is priceless and your message will reach people you wouldn't otherwise target.

Years ago, my friend Madalene proposed an exercise segment for the local TV station in order to promote her business. She has always been a "think outside the box" kind of person. Although her prenatal and post-partum classes were geared to pregnant women and new moms, she worked out skits that anyone could do. Her most memorable segments had universal appeal. She used soup cans for weights instead of purchasing barbels. (Doesn't everyone have a couple of cans of chicken noodle soup in the kitchen cabinet?) For a summer segment, she pushed empty gallon plastic jugs underwater in the pool for working out upper arms. At Christmas she was taped sitting in the middle of a chaotic pile of wrapping paper and gifts and showed how to stretch while wrapping the gifts. And my favorite for all the couch potatoes watching: she created exercises to do while sitting in a Laz-y Boy lounger.

Madalene learned that even a monthly segment took hours of preparation and shooting time. Unless you're a media hog and would really rather be a media personality than a farm wife, plan your PR media events a couple or four times a year.

You don't have to go in front of the camera. Instead you can write stories with recipes and send a blitz of these News Releases to local and regional Food Editors of newspapers or magazines. Suggest a grassfed meal for Valentine's Day. Personally, I think that would be a unique alternative to the same old story about the rising price of a dozen roses (although I wouldn't mind getting the roses!). Other untraditional times and ideas—Quick Fixes for Busy Back to School Days; Fun Fondue with Chunks of Chuck and Grassfed Cheese for the Shortest (Darkest) Day of the Year; Breezy Brisket for the Ides of March. Grab a calendar and think beyond the traditional holidays. You can also go online to find a list of ethnic holidays as an alternative to our Red, White and Blue favorites.

TV and radio interviews. Your first appearance will determine whether the radio or TV station will be open to having you back.

Whether the segment is live or taped, pretend the camera is your best friend. Talk with the interviewer ahead of time so you will know what questions are coming and areas they plan to cover. Be prepared that the host will most likely be clueless. Give them guidelines of questions you can easily answer. Be sure to get in—more than once—how you can be reached or where your product is sold.

Don't wear white, stripes or plaid. Stick with solid colors. Red, blue or black are best.

With radio interviews, the segments are generally longer than TV can spare. Be prepared with plenty of back up information if it becomes necessary to fill time. Talk radio is particularly effective since the audience is already primed for listening to discussions on other subjects.

Be an expert. Although it won't happen overnight, a job well done means they'll start calling you. Concentrate on what you know best—grassfed products and their healthy benefits. When you're comfortable selling grassfed face-to-face to customers, you can easily transfer that expertise to media outlets.

Remind the news director or show host that you can be a "source" at key times—Barbecue Season, Thanksgiving turkey, Easter lamb. Be prepared with a list of tips they can use. The more angles you can give them, the better.

Video. Put together a farm video or power point presentation and keep it running in your farm store. Don't think you can't afford it? Take a tip from the Los Angeles Zoo, as reported by Jay Conrad Levinson in his "Guerrilla Marketing Newsletter." To find a way around a budget crunch, the zoo gave children aged 7-12 a camcorder and told them to go off and shoot whatever they liked. Editing the results, the zoo had seven new TV spots, two of which were in Spanish. Look to your high school or university professors of media classes and suggest the farm-photo-project. You might offer the student a prize in addition to the grade they'll receive.

Most TV commercials only last 30 seconds, some only 15. Keep things short. A three-minute video will be watched in its entirety while one 15 minutes long won't. If you have more to say than can fit into a 3 minute video, break things up by creating another 3 minute video.

News links. The media is always looking for ways to tie a national or regional event to a local news item. Once you've developed a good relationship with a reporter, help them out with "newsy" suggestions. For example, during high fuel prices, a feature on buying local foods (from you) at farmers' markets can save the tremendous fuel costs resulting from coast-to-coast trucking. Plus, customers aren't paying for the truckers, feedlots, or big packers who all take their cut. During a drought be the go-to grassfed guru who can show how pastures tend to stay greener on the MiG side of the fence. High corn prices can be a news peg for all grassfed ventures.

Media master. Create your own media event and invite

Be your own publicist

No matter whether you tap radio or TV outlets, remember, these messages are fleeting. One time just won't do it. By contrast, magazines have the longest "shelf life," especially when they're left in doctors' offices.

the press to attend. This might be a farm tour with a special focus (back to school with school groups), a food tasting festival with a panel of local chefs, or the discovery of a heritage aspect of your farm. Be sure to butter up your reporter with food samples prior to the event.

Be persistent and work to build up your own press.

Advertising

In general, I don't recommend that you spend money on advertising. Any budget you may have for ads is better spent on promotional materials to help you get free publicity. However, there may come a time when paid advertising is justified.

Perhaps you are planning a community-wide festival. Maybe you want to support your child's sports or school activities with an ad on a fence or an ad in a theater program. You might want to pay for an ad as sponsorship of a favorite cause.

Some of the same rules for promotion apply to advertising. Don't talk about you. Talk about them. Stress what your products can do for them. The consumer is only interested in what's in it for herself.

Tests reveal that the only people who "see" your ad are those looking for specific information it contains. You don't see car ads in the newspaper until you're shopping for a car. The rest of the time the ad is invisible to the disinterested customer.

Be aware that the consumer is not as interested in what you are selling as you are. Don't try to be clever or cute. You can't educate your prospects if you don't understand what's important to them and what they want. Your prospect's inertia is your biggest enemy. You need to give them a reason to act, and now.

Guarantee. Jeffrey Lant, author of *Cash Copy*, said the best way to overcome buyer reluctance is with a guarantee. If you are already doing this, advertise the fact.

Make it easy. Lant said people will buy less or not at all unless you make it easy for them to buy. You can do this by including a toll free number, taking orders 24 hours with an answering machine or from your website, and by accepting credit cards. Sure, you will lose a percent of the sale to the credit card company, but the sale is more likely to be larger than if a person pays cash or by check. Tests have shown that credit card privileges will increase the size of the order by 20 percent or more. Some sources say that just by accepting credit cards sales can increase 80 percent.

Free offer. Your advertising should be designed to pull in interested prospects where you can then start a sales program designed to turn them into customers. You should not expect your ads to actually sell the product unless it is less than $20.

What your ads should do is to encourage a potential customer to open a line of communication with you. The best

Narrowing focus

The narrower your focus, the stronger your marketing message will be.

No one wants to buy a product that everyone else has. Your competitors are blasting your customers at the same time as you are with their own advertising messages. In order to be "somebody" to your customer, you have to narrow your focus.

"A specialist can knock off the generalist because the specialist keeps honing their message to the point of being considered an 'expert' in the area."

Think "Grassfed—our specialty—100%."
Jack Trout & Al Ries *Bottom Up Marketing*

way to do this is to offer them something free if they will call you. This freebie should be an information package. This could be a catalogue or a how to, a customer success story, or best of all a free sample.

Put a time limit on the offer to encourage an immediate response.

Show don't tell. Don't have your ads portraying the product. Show the customer what it does. Sell benefits. Not products. Show a picture of healthy children with happy grazing animals. New Zealand grazing expert, Vaughan Jones, said a good ad should be 90 percent green grass, sleek cattle and smiling people and only 10 percent product. The ad should dramatically illustrate a single, unique benefit of the product and only secondarily the product itself.

Do it now. It takes a relatively long time to move a "cold" customer who has never heard of your farm or ranch to the point of being a "hot" prospect. Your marketing efforts need to be cumulative in their effect. Once the sales process starts it must continuous.

For example, when a customer calls, you must have a "bounceback" package ready to mail to them that day. Speed is of the essence as "hot" prospects can quickly cool.

The questions this bounceback package must answer are: 1) Why this product? 2) Why this farm? 3) Why now? The biggest enemy of the marketer is inertia. Lant said a customer who calls for more information really *wants* to buy your product. But if she isn't encouraged to buy it at that exact moment, the odds are something else, or someone else, will get in the way. Lant said the strongest voice to use in your advertising is the command voice: Do this now! People have so many choices to make today, they actually appreciate being told what to do.

Paul Hawken in *Growing a Business* has some good observations on advertising on a limited budget.

Start a relationship. Don't tell your customer everything in one ad. Approach your advertising as though you were having a long-term dialogue with your customer. Let your ads have the expansiveness of a friendship and write copy that respects this relationship. A slower sales approach on the front end usually results in a much longer term relationship with the customer than a harder sale.

Act your size. Small companies should forget full color ads and any other marketing campaigns that feign bigness. If your customer wants big, you can't satisfy him. So don't act big. (Conversely, if you want to create the impression of size and market leadership you must consistantly have the largest ad in your product category.)

Start small. Recognize that your customer probably knows more about your product area and the competition than you do. If you're the new kid on the block, don't beat the customer's door down, let him discover you. Start with small ads in appropriate publications—perhaps a newsletter distributed at a health food store—and keep them there. A small ad becomes like a friend to your customer. It reassures him that your are still in business and waiting to serve him. If your advertising stops, your potential customer usually assumes you have gone out of business. Consistancy of your advertising more than the size of the ad builds confidence in your company and product.

Brochures. Produce a good brochure. Remember, a truly useful product is not difficult to sell. Customers like to discover new things. They then discuss these new discoveries with their peers. Remember, it is this word-of-mouth advertising that is the primary source of actual sales.

While you may like to have a creative series of new ads or brochures each month (or quarter), keep a basic stand-by ad to use when you're not feeling so creative. The worst thing that can happen is not to have your name and toll-free number out there. At the *Stockman Grass Farmer,* we have found that we can run the same ad for several years in a publication with no fall-off in response. However, if we quit advertising the response almost totally quits within two months.

Any follow-up sequence has to be automatic and run by

the calendar. Plan that on such and such a date this will happen and then this will happen and then this will happen. Lay it all out at one time just in case you get distracted down the road.

Adding something new. We've found that when we get excited about a new program, little by little old programs would be dropped, even if they were still working well. Don't discard something that worked well in the past. Add the "new" to it.

Catalog. Do your own advertising with a self-produced catalog. This is an excellent way to expand your direct marketing. You will get far more orders if you put your proven winners in the front pages of your catalog. Intersperse practical tips and information around your products and prices.

What's the cost? Advertising should never cost you a single dime. If the ad doesn't pay for itself in product purchases or a measurable response, something is wrong. Either the offer is too weak or you have placed the wrong product in the wrong media. Do not continue to throw ad dollars after something that's not working.

Talk radio. The problem with advertising is it's like shooting a gun filled with buckshot, hoping something will hit the target. Look for vehicles that have a specific, narrow market audience. Talk radio is a good example. Unlike people who listen to music stations, the listeners of talk radio (and sports stations) are paying attention to what's said. The programing for talk radio has already done much of the work for you by grabbing the audience's attention. Build on that with a catchy ad that keeps their attention. Radio advertising is also less expensive than TV or billboards.

No matter what method you choose, weigh the results by asking your customers, "How did you hear about us?" Then pour more fuel on those fires that burn brightest.

Marketing messages: good and bad

Here are two marketing tales. One lacked marketing thought; the other was loaded with customer benefits.

Waiting for the doctor during a recent checkup, I had nothing to do but stare at the walls. Why do doctors think their patients want to see skinless body parts? You know those posters they proudly display revealing muscles, bones, organs. I'm guessing that even those of you who routinely work with butchers and raw meat don't enjoy seeing human body parts sectioned off like a butcher's guide to rendering a cow, at least not when you're in a doctor's office, or worse, the waiting room. Don't doctors know patients are already uptight? Seeing those posters of what might be wrong only adds anxiety. Or they blow up research about their specialty with words you can't pronounce much less understand. They should give us a children's book to read while we wait if they don't have an ancient copy of *Time* lying around.

As I sat in the exam chair my line of sight went straight to one of those cutaway body part pictures. The light reflected off the surface revealing scuffy marks as if it had once been used for a drink coaster, making the body part picture look diseased. If anyone had taken the time to look at the room from the patient's (customer) point of view, they would have seen that a little glass cleaner would have gone a long way to show they cared about cleanliness, appearances and their customers.

Then a new technician came in to take care of the preliminary procedures. Although she was friendly, she never said, "Hi, I'm...."

A few days later, I was moving Allan's truck in the yard when I heard a radio spot for my best friend's family auto dealership. It grabbed my attention immediately. I sat awed thinking how many marketing points the brief ad covered. The voice over was done by my friend's son, Breck, who writes the script in his head on the way to the recording studio.

Here's what he said:

"This is Breck Daniell from Daniell Motors in Hattiesburg. Can I take a moment to ask you a question? Why are you listening to my voice at this moment as you are going about your day? Talk radio is about wanting to be informed, educated, and productive even as we go about our daily routines. At Daniell Motors we believe you as super-talk listeners are the kind of customers who would enjoy shopping with us. As informed people we all have the means to access the resources to study what we need as well as what we want in our next pre-owned automobile purchase. At Daniell Motors our website is updated daily at Daniellmotors.com and our goal is to make shopping easy and informative. We are located at 1500 Broadway Drive and work hard to keep a great selection of trucks, cars, vans and SUVs of all makes and models. Our no pressure approach will allow you to look and shop without feeling like we are making the decision on how you spend your money. Give us a call at 264-2210 with any questions you may have and we will do what it takes to earn your trust and your business. Thank you Hattiesburg for 31 years. It is an honor to serve you!" Then his children added the company slogan, "That's Daniell with two 'L's."

Hello. By giving me his and the business name he mentally shook my hand in greeting. Always be the first to introduce yourself to new customers.

Honey sells. Next, he complimented me for being "informed, educated, and productive," because I was listening to talk radio. (Actually Allan was the one who had tuned the radio to that station.) This is the kind of shopper who seeks out healthy food for their families.

Target audience. He benefited by narrowing his target market, eliminating mindless radio listeners of music to reach only the kind of person who selects talk radio. They listen to learn something, just as grassfed consumers do.

Like-minded. He let listeners know there are others like them who shop with his company.

How? He gave three ways to find them—their website "updated daily," the physical address, and the phone number.

What? Without being specific, he listed the kind of products they sell. This gives the ad longevity. Their competitors usually get specific with a particular year and model vehicle currently on their car lot. That means they constantly spend money updating their ads. The Daniells have alternated this ad with another for more than a year. It "has legs."

At ease. The next statement describes the kind of business people they are. "Our no pressure approach will allow you to look and shop without feeling like we are making the decision on how you spend your money." And it shows they respect the customer's spending habits.

Values. They are approachable (no question is too dumb to answer), and value their customers who have come to trust them. After all, first-time shoppers have anxiety over

Start with Talk Radio

As a rule, paid advertising is like hoping the magnet of your grassfed products will draw out those needles in the community haystack. Cost effectiveness is difficult to measure unless you have a specific target. However, if you do have a budget for advertising and want to give it a try, start with talk radio.

purchasing a pre-owned vehicle just as first time purchasers of grassfed meats are uncertain whether they'll like it or not.

Validity. The fact that they've been in business 31 years demonstrates their longevity when other dealerships have gone out of business. Customers of grassfed products want to know you'll be in business long term.

Slogan. Repeating the slogan helps people remember their name. It has become a catchy phrase throughout the community. That's brand identity. The more marketing minded you become, you'll start to see good and bad examples in other businesses in your community, and then you can apply those observations to your own marketing messages about grassfed products. Use Daniell Motors as an example—their marketing (excuse the pun) hits on all cylinders. And it works. Just plug in customer benefits about your grassfed business to make people stop and listen as Breck did.

Can you convert this ad into a grassfed best seller?

This is Breck Daniell from Daniell Motors in Hattiesburg. We at Daniell Motors have had the privilege of providing the Pine Belt with quality pre owned cars, trucks, vans and SUVs for over 32 years. There are many things we value as a company, such as a loyal customer base who knows there will be no surprises or games when purchasing from Daniell Motors. We enjoy being a part of this community we call home, and all of the friendships we have formed over the years. At Daniell Motors our central goal is to help you find what you want and need in an automobile, and provide that product at a fair price with personal service that you can count on. Shop our inventory online at Daniellmotors.com where we update our inventory daily with pictures, detailed descriptions and prices listed on each vehicle so you can do your initial shopping at home. Then call us at 264-2210 or stop by and see our no pressure sales staff at 1500 Broadway Drive at the corner of Lincoln and Broadway. Thanks Pinebelt for 32 years and we look forward to serving your needs for years to come.

Logos

A lot of business people think that they'll wait until their business is earning a profit before they make the effort to look professional. One sign of professionalism is a logo.

Logos help your business stand out and make it more recognizable. Logos lend an air of credibility to your business. If you're in this for the long haul, creating a logo that represents what you stand for should be one of the first investments you make.

Hire a professional. It pays to have your logo designed by a professional, even if you choose to save money by hiring a graphic arts student.

While you can design just about everything these days with a home computer, you don't want your logo to look amateurish. Mark Lilley's school bus comes to mind. He had it painted with stick figures as if children had done the drawing. And he parks it at office complexes so that workers can make healthy food purchases on their breaks. However, a professional sign painter did the work.

Design ideas. Talk with your designer about the message you want to convey with your logo. Give them clippings of logos you like and tell them why those stood out. See if there is a way your logo can reflect your mission statement. Perhaps there's a symbol that represents the name of your farm.

Stick with it. Once you've settled on a design, stick

with it. Any dramatic change you make can cause your customers to wonder if you have gone out of business and been replaced by someone else. That doesn't mean you can't let your logo evolve as your business grows. Over the years we've changed the type face of *The Stockman Grass Farmer* to reflect our evolution as a cutting edge publication. The words are the same, but we added color as we "grew up." If your logo changes, use the change as part of your marketing message. Incorporate the words "Now" or "New" as you draw attention to the changes with your logo.

Appearances. A well designed logo can make your business look bigger than it is. Anything that's home-printed using available clip art will look amateurish and small. The logo helps establish your business as one with a track record.

Branding. A logo is a symbol of your branding. It represents commitment and longevity. This is an unspoken means of gaining the trust of your customers.

Memories. Prospective customers may not remember

Designing logos

1. Think long term with the design. Avoid trends so that the same design will be recognizable and valid 20 years from now.

2. Keep it simple and easy to understand. This is where the talents of a professional designer come into play.

3. Be sure it's unique and doesn't hint at anyone else's logo.

4. Does it symbolize your name, your business, your identity, your future plans?

5. Make it recognizable regardless of the size.

6. Use it everywhere.

Jay Conrad Levinson, *Guerrilla Marketing*

your name, but the visual graphic of a logo will aid recall. Capitalize on this by making all of your marketing materials consistent. Put your logo on everything from invoices to outdoor signs.

Stand out. A logo will make you stand out from the competition who hasn't taken the time to have a logo of their own.

Brochures

Once you have a logo, create a brochure. This then becomes the building block upon which you can later add other marketing materials such as a catalog or newsletter. By creating the brochure first you can borrow elements of it to give the catalog or newsletter a unified look. Use it as an identifying feature for your farm. Think of it as if you were dressing all your children in a tartan plaid.

Why? People equate brochures with "real" businesses. They like to have something to take home and read at leisure. It gives them an opportunity to tell others about what you do.

Spread the word. Brochures are not intimidating. They can be placed in like-minded businesses where curious people are likely to pick them up. Everyone loves freebies. Because you'll want to use them in many places, design it in such a way so that you can print a lot of them without feeling guilty about the cost.

When you ask another business person if you can leave some of your brochures with them, track the effectiveness of the location by how frequently you restock. Find an appropriate holder to display them. Something that holds the brochures upright will make them easier to see and take up less space rather than just leaving a stack sitting on a counter.

Post them on church and college bulletin boards. (Students may not have the funds for your products, but their professors will.) Leave them at restaurants, health clubs, health food stores, cooking stores, veterinarians, doctors' offices. Ask every friend you know if you can put them in their business office. Always ask permission first before leaving them, and don't expect anyone to police them in your absence.

Front. Use an enticing picture with an attention-grabbing headline on the front. Some direct marketers feel the cover should have one or two highlights of the content.

Inside. With a catalog or newsletter you want the content to look busy. Brochures are the place to use white space for emphasis. Use the inside front to make a bold, standout summary of key points you want the consumer to know about you and/or your products.

Every page should have a call to action: Call us at....Visit our website at....

Text. Think of your customer as you prepare your text. What do they need to know? How can you help them? Write it as if you're talking to a friend. Avoid technical jargon. Be folksy, however, stick to the benefits.

Q&A is an attention getting format. People read the question and then want to know the answer.

List benefits of your products—100% grassfed, never fed any grain, no antibiotics, no growth hormones, healthy omega-3s, CLA. If there's room, list your products and prices. If there's not room invite people to ask for a free catalog.

Brochure Check List

1. Have you put the customer first?
2. Have you included the benefits of your product?
3. Have you given the customer a reason to act now?
4. Have you listed all the ways to contact you?

Keepers. Including a how-to message or grassfed cooking tips can give the customer a reason to keep the brochure, which makes it a reminder of your products every time they use it.

Photos. Pictures will have more impact if they relate to the message you want to convey about the farm and your products. Always include identifying captions. Don't overlook the whimsical use of cartoon blurbs posted above someone's head.

The Parry family of Fox Fire Farm in Ignacio, Colorado, used the first generation family photographs as shadow-background on the pages of their 21st century brochure copy and photos.

Names. It's particularly important in a brochure to identify yourself. Include a picture of your family with everyone identified. Make it folksy—Abbie handles our accounting, Bob works with the beef cattle, Claude raises the chickens. At the very least put someone's name with the contact information on the last page. This adds to your brand identity and the authenticity of your working farm family. People want to know who they are dealing with as well as the name of the farm.

The back. A simple tri-fold brochure is the simplest to begin with. Plan the content in such a way that the back third can be torn off and used as an order blank.

Make sure you include all of your contact information—email, phone, farmers' markets dates and location, farm store—to make it easy for prospects to find you and place an order.

Take action. Give them a reason to take the next step. Ask them to call, write or email for a free catalog (which will list all of your products and prices). Do it now.

Catalogs

Catalogs can help increase sales and capture new customers. Although they differ from brochures and newsletters, they may contain common elements for effectiveness. Once you create a brochure, you can borrow features from the brochure or newsletter to design the catalog. Perhaps your catalog will be a hybrid of a newsletter—a news-a-log or a cataletter—with a mixture of news items and products. Both can be used for marketing products and educating consumers.

The main difference between a catalog and a newsletter is that the catalog will have a timelessness and longer "shelf life" until a new product needs to be added to it whereas the newsletter should have fresh content each issue.

A "NEW!" catalog sent to old customers will out pull a mailing to non-customers by 400 to 800 percent.

Research. As a freelance writer, the best way to sell an idea to an editor is to study their publication cover to cover. What kind of articles do they include? What is the style? Who are their advertisers? What do their readers want?

The same careful study applies here. Begin by studying the catalogs of your competitors.

1. Who are their customers? Who are you targeting?

2. What are they offering? Are their products strong in one area—beef or lamb?—and weak in another—eggs or cheese? Can you fill a niche they don't have—pet products—

or personal products such as goat milk soaps?

3. How are their products priced? Does that differ from your pricing philosophy? What market does that appeal to?

4. Do they offer seasonal items? When?

5. Where can their products be purchased—on line, farmers' market, farm store, buying clubs? What about yours?

6. What do they do best? What are their weak points? Can you do better? What if a prospect received your catalog and theirs the same day, which would they order from?

7. Why have a catalog? It can show prospects the multiple choices you offer. And it becomes another marketing tool in your arsenal.

Readable. While it looks good to have a "busy" page with lots of products, be sure you haven't reduced the type size too much to get everything in. 12 point type is best, 11 is also good. Smaller type can help save the number of pages to print and cut down on weight and postage if the catalog will be mailed. However, if you have a lot of senior customers, consider using 14 point type as a benefit to them.

Be sure your prices match the description or photo of the item. Use lines if necessary to separate items so there won't be any confusion over which item costs what.

Wow factor. If your budget allows, professionally designed, four-color, slick paper catalogs will surely wow your prospects. However, the content can be equally impressive in black and white. In fact, a simple, homespun appearance may fit your goals best. If your catalogs have a high price per unit, you will be less inclined to give them away.

Use pictures if possible, even in black and white. Rather than photograph meat or cheese, use farm photos to reinforce the message of who you are. Focus your **wow** attention on the cover or envelope if they are to be mailed so that they will rise above the regular junk mail.

If in doubt use more text description rather than a photo. A folksy, chatty style can make up for the lack of slick paper, color pictures or expensive artwork.

Covers. There are two schools of thought about front covers—one to use it to establish a mood, or two to use it to sell something. Study other catalog covers and decide which ones appeal to you. Place your logo somewhere on the cover, even if it's positioned in a bottom corner. This keys off your brand identity and should be a part of every catalog, newsletter and brochure you produce.

Back covers are where address space appears if you plan to mail the catalogs. Because of this it's often the first thing a person sees, which results in it being the catalog's best-selling page. If you will be mailing in bulk, reserve space for postal permit information. This is a box in the stamp area that reads, "Bulk Rate, U.S. Postage Paid" with a location, company name or permit number. Check with your post office for details. To serve double duty for the accuracy of your database of names, print "Address correction requested" in the bottom left hand corner.

Text. When you are writing descriptions for various products you can sometimes repeat statements. People don't generally read catalogs page-by-page the way they read a book. Instead, they jump around looking for the items that interest them. While you will need to include distinguishing characteristics for each of your products, you don't have to create a soliloquy for each one. Take advantage of this opportunity to educate your customers. Answer as many questions as you can think of with your text, otherwise whoever answers the phone

How passionate is your customer?

"Where to shop is based on a consumer's passion about a product. If there's no particular passion—white socks or underwear—the consumer will look for the cheapest brands. Passion inspires the customer to seek the best."
Laura Cutler *Inc Magazine*

or receives the email order will have to do the work your catalog should have done in the first place.

Creativity. Don't get so caught up in being artsy to the detriment of your products. If you picture a nice steak on the grill or a fancy platter, be sure you customer knows you're not selling barbecue grills or dinnerware.

Extra! Extra! You want people to read all about it, so don't ration them out. The more places you can distribute them, the better. Readily give them to satisfied customers and ask them to pass them along to their friends. Keep extra copies in the farm store or at your farmers' market booth. Drop them in the bag with every purchase.

Benefits. Since a catalog is another marketing venue, use it to stress the benefits to your customers for purchasing your products. Unless you're going for a high retail, sophisticated design utilizing a lot of white space, take advantage of those empty "holes" by including recipes, cooking or shopping tips, an explanation of different named cuts of meat, anything that benefits the customer.

Design. Even if you plan to use a commercial graphic designer for your catalog, prepare as much in advance as possible before you turn over the materials and project to them. Just as film makers start with cartoon "story boards" to plan the scenes and action in films, sketch out the pages with all those things you want to include from your logo to your mission statement. Making this rough draft will help you see if you need more or less pages. Think in terms of multiples. For example, if you take a regular sheet of 8.5 x 11 copy paper and fold it in half you will have 4 digest size pages. When you add another sheet you'll have 8 pages (front and back). You will get a higher response from a 32-page catalog than a 24-page one. The longer you keep someone reading your catalog copy, the better your chances for success.

Easy as meat pie. Make it easy for your customers to buy from you by placing your contact information—phone, address, email—on every page. Somewhere put a person's name with the business contact information. The personal touch means prospects won't be dealing with a commodity style operation.

List all pertinent information—weights, sizes, names—on the order blank so that all the customer has to do is check off the item rather than flipping back and forth between the order blank and its description on another page.

Reserve space on the order page to ask for names and addresses of friends who'd like to receive the catalog.

Unless you go to the expense of having the catalog printed by a professional printing company, convert it to a PDF document that can be emailed to prospects. Getting an immediate response to their catalog request is impressive. It's easy to do. Just ask a computer guru kid to show you how.

Tracking. Now that the catalog is mailed or distributed, track sales page by page. This way you can determine your best sellers and put more fire in that engine next time.

Newsletters

Newsletters are a way to have a regular conversation with your customers. Newsletters keep them informed about advance items non-customers don't know about. If you include beneficial information it can add to their education about grassfed products. And it can be fun.

Stumped on how to start? Don't begin your newsletter with "Dear Friend," or "Dear Valued Customer." Write a catchy headline. Spend some time thinking about what you'll write long before you begin. Keep a file where you can drop in notes in between newsletters on ideas for the next issue.

Content. The same things that apply to catalogs can work for newsletters. However, newsletters need more editorial content. Keep your customers informed about farm happenings —the new calves or lambs, addition to the farm store. Let them know what's currently available and what's coming, such as holiday turkeys. Promote your events.

Ask other family members to write sections to keep the content diverse. Add "words of wisdom" including quotes you like or something funny or lighthearted. Include excerpts about grassfed seen in other news media. Give your readers insights into your family history including marriages, new babies, significant anniversaries. Short bites of information work better than a long novel approach. Use subheads to break up text, or box certain paragraphs to draw extra attention.

By tracking the effectiveness of your newsletter's content, when you find something that works one year, recycle it the next. Few people will remember it's the same.

Customer contributions. Get your customers involved. Ask them to share a recipe or tell a story about how they used your products. Feedback may be more forthcoming if you ask them this in person when they shop with you. Some people just won't take the time to respond to a polite request. However, you could encourage participation by creating a contest for the best story or recipe.

Interviews and profiles. Write an interview or profile of your family members who are involved in the farm. What would your customers want to know in order to feel like you're their best friend?

Take action. When Joel Salatin started sending out his annual newsletter it asked people to call. Few did. They tried an order-blank mailer, but that wasn't effective either. They subsequently learned that including a self-addressed envelope spurs orders. To spur their customers to action, they tell everyone orders are taken on a first-come, first-served basis.

As orders are received, they are dated. Advance ordering data helps them judge whether they need to add another batch of chicks or call in a like-minded producer to help fill grassfed beef orders. The Salatins also keep a "safety valve" list. They refer to this whenever there's a no-show or someone is willing to increase their order if there is excess product.

Talk to her

In writing a newsletter, pretend you are having a long conversation with your customer. Remember that one sale to a new customer will take six times more time and effort than making a second sale to an existing customer.
Paul Hawken *Growing a Business*

Be transparent. Salatin suggests keeping customers informed to gain their trust. "If you had some tough meat, explain why and what you're doing to fix it," he suggested. "If you are increasing prices, explain why."

Orders. Salatin's order blank lists all of their products within specific time frames. Beside each batch of chickens is a date they will be ready so that customers can check not only what they want but when.

Internet marketing

"A website is your store front to the world," said Joel Huesby of Thundering Hooves Ranch in Walla Walla, Washington.

Having a website allows you to refer your customers there to learn more about you. By setting up a system to take orders over the Internet you open up one more venue for marketing. Making it easy for your computer-savvy customers to buy from you shows you care.

Benefits of Internet marketing:

Time saver...for you and your customer. It's easier than making a phone call or filling out a catalog's order blank. Since web pages are set up to allow customers short bites of information (with links for lengthier discussions or more details), this respects your customer's time by allowing them to click on just the amount of information they need to satisfy their curiosity.

It saves time for you by not having to constantly explain the benefits of grassfed products, directions to your farm, or who you are. And, orders can be placed 24/7—at the convenience of the customer, who may be wearing hair curlers and a bathrobe. That allows you to process those orders any time you choose...when you might be still in curlers and pajamas. Even online, women will be your main customers. Anything you do to save their time is priceless.

Say more. While each web page should be kept short, it can be linked to other pages to tell more of your story and say more with multiple short bites rather than one long, rambling message. Remember those books you read as a child with a flashlight under the bedcovers, promising to quit after just one more chapter? But the chapter left you guessing, so that you had to read just one more. Think of your website like a mystery novel. When the novel's chapter (or web page) ends, is there something that is left unexplained to urge your reader to keep reading (or clicking)?

Keep information bites short and crisp with plenty of click-on options. Think of the website as if it were a children's pop up book. Each page has links that lead to the next pop up page. This will help you to organize everything you want to put on it. Get a notebook and collect all the pages and their links before you turn it over to a designer or that computer-talented family member who can produce it "on-farm."

Cross marketing. Web marketing can be used for up-marketing—getting your existing customers to buy more or pay more for something. For example, Peapod.com is an online grocery store. When a customer clicks on hotdogs, up pops complementary items—buns, mustard, relish, napkins. They service metropolitan areas such as Chicago, Boston, Southern Connecticut, Long Island and Washington, D.C., with home delivery the day after the order is placed.

Some of Peapod's shopping options include Item Comment—the customer tells Peapod how they want their bananas, ripe or green. If an item is unavailable, Peapod suggests substitutions. Customers can shop by Previous Order, a shortcut that pulls up the last list ordered. (Don't most of us tend to buy the same basic items each week?) Sort features allow customers to group items by price, nutritional content or even "most popular." Aisle Browsing on site has icons to represent a virtual grocery aisle. Customers click on the aisle they want to use. Keyword Search allows customers to type in a specific brand name. For delivery, customers choose At Home or Unattended Delivery. Shopping and delivery service fees range from $2.95 to under $10.

Younger, more informed customers...are more computer and Internet savvy. The first place they turn when they want to know more about something is an Internet search engine, like Google. It's a luxury to have customers looking for...and finding you via the Internet instead of you constantly hunting for that needle-in-a-haystack customer. Customers who are *looking for you* will make the best repeat customers.

More and more producers are realizing that Internet sales are making up a hefty segment of their direct marketing dollars. Bite that bullet and get on the Internet.

Make a bold statement. Those of you who have resisted direct marketing because you're shy or feel you aren't a "people-person," can use your web site to do the talking for you. Just be sure it's as bold and beautiful as what you produce. Grassfed products that are "just good enough" will harm you, other grassfed producers, and the grassfed industry.

Updates. Once the website is up and running, updates (price changes, product additions, seasonal recipes) can easily be done. Even if you hire a web designer initially, they should teach you to make these changes yourself.

Pricing. Here's some good news. A study by MIT revealed that few consumers expect to find *lower* prices on the Internet. A wider choice of rare and unusual items is what attracts them. Many Internet customers are high income families who want to buy from people who appear to have similar values.

Validation. When your products are mentioned in the press, you can list a link to the source. Links to sites such as eatwild.com or localharvest.com also give your products validity. Include rave reviews from customers and success stories about using your products. Most importantly, in today's marketplace, warned Mart Martin, creative director of Jackson Spalding in Atlanta, if you are not on the web, you and your products are suspect.

Things you should do:

Make it easy to order. Make the process of ordering so easy a 10-year-old can do it. Well, ok, today's computer-savvy kids already know more about computers than you do, so make it easy enough for a 60-year-old to figure out. Not long ago it took me three and a half hours to file and pay for a copyright online. The outdated paper version (no longer available) took five minutes. Most of the my time was spent figuring out how to give them my money online.

Respond quickly. Respond by the next morning at the latest. Even better, set up a system so that the customer immediately receives email acknowledgment.

Inventory. Before venturing online, be sure you have the capacity to profitably produce enough inventory to supply your online customers.

Research. Check out your competition online. Study what makes you and your products unique in comparison. What are they doing right and how can you top that? What can you add that they haven't thought of? Using a friend to insure your anonymity, order something from their website to judge their response and professionalism.

Photos. People want to know the people they buy from. Include photos of your family, your children, and your pets, with photos of your farm. If at all possible, have one or more of these photos on your home page. Introduce yourself before your customer begins shopping.

The typical web shopper will decide within two to three seconds if they are attracted enough to the site to keep reading. Homepage turn-offs can be too much text or too many photos. Alderspring Ranch in Idaho has six photos placed in a zig-zag, overlapping position as part of the banner, which makes them available for further scrutiny without being a distraction for someone looking for "just the facts, ma'am." They use sidebars of information with photos as brackets on both sides of the central text.

The photos you select are key. Consumers want to see

those shiny animals grazing lush, green pastures. They should convey authenticity and the cleanliness of your farm. Pictures of the producer and family show that this is a real working ranch, not a corporate stand-in.

If you don't have any children, borrow some—nieces, nephews, grandchildren, neighbors, customers. Happy, healthy children interacting with your animals or products score points with consumers. Do you have any "aahhhh" photos that might melt her heart?

Remember, women are connectors. Photos give her clues that there are other customers just like her shopping on your website. The appearance of your website should reflect the world that your customers live in to make them feel at home.

For example, if your target market is upscale, professional women, photograph your products served on classy china place settings or an attractive, colorful serving platter placed on a granite kitchen counter top. Perhaps you're promoting poultry or picnic items. Set an old-fashioned quilt in the pasture with a handmade picnic basket and red-checked napkins, and let your fried chicken be the star in this scene. Maybe you want both on your website to appeal to diverse audiences.

Stealing home. According to Jo Robinson of eatwild.com, "The purpose of your home page should be to let the customer know you are there to sell grassfed products. Make it clear how they can do this. A toll-free number is attractive."

Name names. One Christmas we ordered what was advertized as a grassfed turkey through our fledgling organic food store. We assumed we were getting a genuine bird from a regional, if not local producer. When we picked it up, the label told us it had been shipped from California. And when we looked up the company online we couldn't find anyone's name, no mention of how the turkey had been raised, and no pictures to reassure us we had received the grassfed turkey we dreamed of ordering. Lack of clarity on the website made us feel as if we'd paid a fortune for something little better than the supermarket variety. We came to the conclusion it was a corporate bird, and our money would have been better spent online ordering a turkey from a grassfed producer's website thousands of miles closer to home.

Like photos, having your name on your website makes your products real to the consumer.

Benefits. Turn everything you are doing into a consumer benefit. For example, you can state, "USDA inspected *for your safety.*" Meat safety is the number one concern of mail-order buyers.

A consumer survey found that 70% of Internet consumers would pay a premium for a product with the claim of "Raised on our farm from birth until market."

Another key concern is the humane treatment of animals. Apply for a "humane" label from one of the established Animal Welfare groups. This is particularly important if you are marketing veal or poultry.

A claim of "grazed on 100% Certified Organic pastures" has more impact than just a Certified Organic label. Consumers are learning that many organic meat products can come from industrial farms. They want to be reassured that your claim is real.

Guarantee. Because people are reluctant to purchase something sight unseen, it's important to offer some kind of guarantee in order to gain their trust. Money back is one option, but better still is to offer credit toward the purchase of *something else.*

Take credit cards and offer shipping. While you can specify that payment can be by cash or check only, sales beyond your local area will work most efficiently if you take credit cards. Internet shoppers expect to pay with plastic.

Some grass farmers with websites (Joel Salatin is one of them) will not ship. Salatin feels his sales should come from a "local" area he defines as within a few hours' driving distance of the farm. He's not hurting for sales because of strict adherence to that principle. In fact, he credits the fact that he refused to ship his grassfed products to Michael Pollan as to the reason Pollan came to Polyface Farm and prominently featured Joel in his best seller *The Omnivore's Dilemma.* Joel maintains a website for credibility, and to help his customers learn more about grassfed.

Internet tools

Using an Internet database allows you to communicate with your customer. Those who sign up for your web "news" are giving you permission to reach them. There are a couple of things to remember. They don't care about *you.* They care about *me.* Your messages have to relate to their life.

As Seth Godin pointed out in *Meatball Sundae*, "The moment the messages you send me cease to be anticipated, personal, and relevant, then you cease to exist in my world."

Blogs:

Should you add a blog button to your website?

Do you have something to say? Blogs are people connectors, but they're only read if the blogger has something to say, and is willing to keep the blog going with regular posts. What you say has to be of interest to your readers.

The most powerful example of this for me is what happened to a young dancer named Nicole, the 24-year-old daughter of one of my girlfriends. Nicole fell six stories from her apartment building in New York on Labor Day weekend 2008. She lay in the air shaft eight hours before the janitor noticed tennis shoes with her body attached and got her to a hospital. Her parents were told she had less than a 50% chance for recovery to which they replied, "We'll take that chance."

Within days, her mother started a daily blog on caringbridge.com, a fabulous website that allows friends and family to keep up with the long-term progress of someone in the hospital, post encouraging guest messages, see photos or videos and support the family and patient. The site allowed her family to keep everyone up to date on her progress without having to repeat themselves endlessly to those who shared their concern. For the first six months of her recovery, the caringbridge site was the first thing I checked every day and I still log on now that Nicole is typing the blogs herself.

Nicole is a walking, dancing miracle. Within six months of the accident, this dancer was in a specially choreographed stage recital. Even earlier she began giving inspirational speeches to help others. Now she is re-learning how to drive a car as she continues to regain the full use of her hands (which with her multiple spinal injuries is one of the last functions to return).

Tell the truth. Blogs inherently require honesty.

It's not about you. Remember, if you blog, the content has to be something your readers care about. On stockmangrassfarmer.com Allan posts newsy items on his blog. The immediacy of many of these items would be sadly out of date if they were held for the regular monthly printed magazine. And, many of these blog bites would never make it into the magazine due to lack of space.

Readers who regularly check Allan's blog will access other pages of the website and can often be the first to learn about new, upcoming events and new book products.

Press releases:

Whenever you have something new—new baby, mention in the press, additional product or service—post a press release about it on your website. "You do not necessarily see a huge increase in sales due to press releases," stated Brent Walter of Fox Fire Farms in Ignacio, Colorado. "However, you get more traffic and more popularity with the search engines, which will hopefully translate into future leads and sales.

Surveys:

Women love surveys and quizzes—just check your favorite women's magazine as proof. Quizzes can help you collect lifestyle info. Using your web database, you can set up surveys for customer feedback.

In *Marketing to Women*, Marti Barletta wrote that men are more interested in what *he* thinks about something. Women are more interested in how *others* think about something.

Clarity. Be specific with your questions. For instance, don't ask "How important is quality to you on a scale of one to five?" Instead write "How do you measure quality? Please rank...tenderness, taste, appearance (marbling, fat content), shrinkage when cooking, sales service."

Why survey? Only create a survey when you want to obtain useful information to help improve your product or discover weak links in your service. This can help you determine how close you are to producing and selling a truly quality gourmet product. From the responses you can create your own research study to validate who you are and the services you offer.

Be specific. Don't ask your customer to rank the top three out of five choices, or ask how important something is—very, somewhat, not important. What does that tell you? Not much.

Be specific using a case in point:

If a particular cut of meat costs 20% more than you planned to spend, rank in order the reasons you might buy it anyway:

Worth it—artisan-produced on a small scale (not from mass or factory production).

Healthy—grassfed, CLA and omega-3 rich.

Local—knowledge of the farm or farmer who produced it.

Environmentally friendly—farm has clean water, green, growing pastures.

Pre sold—because a friend recommended it, or they have purchased before.

Too busy to shop elsewhere—message clear and professional, to the point.

Other—let the customer fill in the blank. Always

provide ample space for write-ins.

Finally. Be sure to share the results—online.

Links

Search engine optimization. "A well-linked and optimized website, equipped with relevant and quality content, will give you the ability to drive sufficient local web traffic to your site," explained grassfed marketer Blaine Hitzfield in *The Stockman Grass Farmer.* "With proper Search Engine Optimization you will also have the ability to target specific nearby cities and metropolitan areas...(from your rural farm location)."

Links. When people do a Google, Yahoo, or other Internet search, very few look beyond the first page that pops up. The key to being in the top ten is having other websites that link to yours. Eatwild.com is the first listing after the advertising links and should be the first place to which you should link your website. Jo Robinson's site is the pioneer omnibus site for grassfed products. Eatwild contains the latest grassfed health news. For an annual fee of $25 Jo will add your link to the list of producers by state. This simplifies things for customers by narrowing their geographic search.

How do you find more like-minded customers? Add a click to *Tell a Friend.*

Invoices and order forms

Everything you do should reflect the quality of your products, right down to the invoices and order forms you give customers. Once you've made the sale, don't think of these documents as an afterthought. Instead, use them as another vehicle in your marketing strategy.

Who are you? Print your logo, company name and contact information on every form. People who buy from you infrequently may want to refer back to the form at a later date to contact you. They can use the form's information to tell friends how to buy from you. And it just looks more professional than using a generic receipt book and a rubber stamp. *The Stockman Grass Farmer* has double-sheet order forms that allow an imprint of the order on the second sheet (with a duplicate of our contact information). This allows the customer to have a receipt while we keep a record of the transaction. The top of the form below our contact information has lines to write the customer's namc, address, phone, and email. "Credit card...Check...Cash" can be circled to track the nature of the payment.

Validity. Print a "statement" on the forms to show the validity of your business. Perhaps it's your farm mission or slogan. Maybe it's letting customers know you're not a new business. If you write the number of years you've been in business you date the usefulness of the form. Instead write something such as, "Serving health-conscious customers since 1990."

Make it personal. You've started a relationship with the customer with the sale. Make it genuine by adding a personal handwritten note. "Thanks for shopping with us." "Let us know how we can help you again." "Enjoy!"

Be friendly. If you haven't been paid in advance (and

Statement-based marketing

Using your statements and invoices as marketing tools gives you a competitive advantage. People can be bombarded by 3000 advertising messages a day, yet they will receive fewer than 12 "transaction" mailings a month. That makes those mailings invaluable for building stronger customer relationships.

People will open bills and statements containing financial information. Ordinary letters may be tossed unopened.

The average person reviews a bill for two to three minutes—longer than most TV commercials.

Invoices and statements will usually be viewed multiple times—first to check for accuracy, second when payment is prepared, third to submit as an expense for reinbursement (if this situation applies), and fourth to file the document.

"The rewards of statement-based marketing couldn't be greater—increased satisfaction and customer retention, improved loyalty and corss-sell, and greater profits from customers."
Suresh Nair, director of strategy for Pitney Bowes Business Insight

you should have been), write a friendly reminder such as, "We appreciate your promptness in payment."

My best friend, Madalene, was thinking about changing the company's insurance to a local firm. After she got the quote from the new company and studied it, she noticed discrepancies in the coverage their current company offered. She did some comparison research and asked more questions of the new local firm. The local agent sent her an invoice with a note berating her for comparison shopping, and saying if Madalene had given them the quote from her current insurance company the local firm would give her a better deal. Shouldn't the local firm have given her the best possible pricing in the first place, particularly since she was a new prospect and they needed her business? Would anyone choose to work with a company that demonstrated an adversarial attitude before the purchase was made? What kind of relationship did they plan on building? Madalene chose to stay with her current insurance carrier, even if it cost her a little more.

Promotion. Use the forms to promote something new, different, expensive. "Ask about our gift baskets." "Treat your best friend to a Gift Certificate. What better present than a home-cooked meal?"

Double duty. Depending on the design, the invoice or order form can double as a tear-off "keeper" brochure that can be given to friends. Or the fold-over segment of a tri-fold form can be used to conduct a customer survey "to better serve you." Maybe it's an ad for upcoming specials. The only time you should leave one side blank (except for your logo and contact info) is if you plan to use the order form as a mailer.

Easy does it. Make it easy to order with a form that lists all of your products, weights, prices so that all they have to do is check the item or carry the total amount to a space on the same line at the right. When you send an invoice, include a self-addressed envelope for a prompt reply. Adding a stamp makes it even easier to send a check, now.

Use that blank space. If you can't think of anything

you need to promote, use the blank space to include a recipe, grassfed benefits, news of an upcoming farm event, tips on how to stretch food dollars, shopping tips, or notices of special items such as holiday turkeys. If you've printed too many forms to make it costly to add additional "news" print the extra information on labels and stick them over the blank space.

Guarantee. If you guarantee what you sell, be sure your customers know about it. Include that information, don't just assume they know.

Blurbs. Include a request for customer referrals.

Compiling a Farm Manual

As a multi-tasking female, do you make lists so that you get everything done that needs to be accomplished?

Are you constantly making notes or pulling magazine clippings of things you want to remember or refer back to some day...and then forgetting where you put them?

Maybe it's time to create a Farm Manual. Think of it as a diary for your farm or ranch. Buy a thick-spined three-ring binder and a hole puncher. You might also want a package of "Table of Contents Dividers." These come in various sizes from 4 or 5 to 32. First, you might want to create your own if you don't know how many categories your manual will cover.

At our church, we ask each ministry leader to create and periodically update a Ministry Manual that we file in the church library. This not only contains a history of the ministry's development, procedures, activities, and accomplishments, but becomes an invaluable guide when there's a change of leadership. It's a quick start aid for the new leader.

What should you include in your Farm Manual?

History. Maybe you've told your children the history of the farm, but have you ever written it down? After all, some of the facts might get lost or confused if yours is a multi-generational farm. Or maybe the newcomers—your children's spouses —will be interested in knowing the farm's history. Make it as elaborate as you like with early farm and family photos. Add a time line of when the family started farming or ranching, when the land was purchased, significant dates in its development. With all the history in one place, the Farm Manual makes this easy to reference when you need it.

Progress record. Only when you write things down, and keep records, can you clearly see how far you've come. This sets the stage for what needs to be done next. This is a history in itself. Categories you might consider are Pasture Layout; Pasture Inventory Records; Minutes of Family Meetings; Promotional or Marketing Ideas.

Pasture designs. Keep sketches of paddock subdivisions here. Make notes of what worked (or didn't), when and why. Maybe this section should contain a series of dated photographs.

Writing the Farm Manual

What's the purpose of the manual? This will help determine if it needs to be fancy or can be folksy, a place to collect items in pockets or more organized in a 3-ring binder.

Who will be using this? This will determine the durability of the manual or the number of copies that should be made.

What should be left out? If non-family employees will be using or adding to the Farm Manual you won't want to include financial data. Keep that information elsewhere.

Where should you begin? Anywhere. Worry about the organization (alfa by topic or chronological) later.

How will you find things later? Tag each section with a title as if it were a chapter in a book. List these "chapters" in the front as if it were an outline.

Organization. I'm a packrat. If I think something is important, I'll make a file and box it away...and usually forget it. Needing more shelf space for *Stockman Grass Farmer* issues, I tackled the boxes of disorganized materials stuffed into our storage office. I shredded ancient bank records—so many I broke the shredder and had to buy a new one (this one can even shred expired credit cards!). I spent days organizing and labeling photos and negatives and 35mm slides. I even found a tube of Chapstick from the 1970s. (Why did anyone keep that?) Amazingly, Allan helped. I told him there were train photos mixed in with everything! He kept saying, "I thought these were lost." Best of all, in my SGF box, I found old files of things I'd forgotten. With a little updating, I could use many of those old ideas for "new" book promotional themes, and even topics for my column. It was a gold mine of good content just gathering dust.

How to. If there's something you don't do often, yet need to remember how to do, log it into the Manual. Perhaps it's things to remember about hiring an employee. Maybe it's a list of market rules for various farmers' markets. It could even be things to remember next time you have a combative customer.

Procedures. It is particularly important to create procedural documents if you are working with interns, employees, or family members unfamiliar with the tasks you assign them that you are generally in charge of. This might include how to sort and pack freezers for buying club deliveries or the farmers' market. Perhaps it's how to set up paddocks and where to place gates—and why such choices are being made in a particular manner.

Promotional ideas. Keep a copy of every brochure, recipes, signage content that you use for promotional material. The manual makes it easy to pull out an old idea that worked when you're creatively blocked about creating something new.

Ideas. This section can contain those magazine clippings you want to follow up, or random thoughts that appear to be good ideas you don't have time for now. Write these down and keep them together for snowy winter evenings or more leisurely times.

Resources. I put together a manual for my mother's new house. It had a list of key people and how to contact them: the contractor, the electrician, the appliance dealer, the security company, the water company, heat and air contractor, plus a list of phone numbers of her neighbors. I punched holes in the appliance operating manuals too. Everything from the dish-washer to the gas fireplace is in one place. For easy reference, list vendors or contractors for your farm operations.

Copycats and niches

Everyone says imitation is the sincerest form of flattery. It's also a pain when your competition starts copying you.

What should you do when your competition "flatters" you by becoming a copy cat?

Nothing. Realize that no one can completely duplicate what you do because of who you are. Your personality should rub off on your marketing messages. As long as you stay true to yourself and what you believe, the uniqueness of your position can't be duplicated. Accept the copycat as a compliment.

Investigate. Take a serious, objective look at the similarities of your competition. Be honest. Perhaps they've found a chink in your armor and have picked up on something you've missed. Make note of things that are the same and the differences. How can you improve on something you're both doing?

Be unique. The more unique you and your product are, the more difficult it is for anyone to duplicate. Unless you're a major player like McDonald's where thousands of hours and even more dollars are spent making certain every franchise does everything exactly the same, no one can duplicate your unique position.

Copy a copycat. Only those in your immediate marketing area pose a real threat. Your closest ally can be a copycat who lives 1000 miles away. Network together to share tactics and information on what works and what doesn't.

Flexibility. Give up your comfortable, rigid thinking and procedures. Now is the time to take a risk and try something different or add something new.

Service. An intangible thing that is difficult to copy is how you treat your customers. The more you can build a personal relationship with them, the less you have to worry about them shopping elsewhere. As Bo Burlingham states in *Small Giants*, transform your marketing from a customer-friendly business to a customer-intimate one.

Carving out a niche. Finding a niche in business means setting yourself apart from your competition. At least for awhile. Once competitors figure out what you're successfully doing they will try to copy you. Keep dancing. Keep changing. Niche marketers are pioneers who like to be on the front lines all by themselves. When others move in they pull up stakes and find a new wilderness where they can be alone again.

"The niche you choose will drive your marketing," said Jay Conrad Levinson, author of the *Guerrilla Marketing* series. To find your niche, focus on your competition, your target audience, and your own capabilities.

Be the one. Can you be the only....? Can you offer a service no one else has thought of? How quick is your delivery service? Do you service the only metro buying club in a particular area? Do you have the widest selection? Are all of your products local?

Be unique. Every piece of communication a niche

Niche markets step-by-step

1. Look for a market with no competition.
2. Be as different as possible.
3. Communicate that difference.
4. Concentrate on profits rather than gross sales.

marketer produces must scream, "I am different!" That difference has to be obvious to the consumer. In a niche market you're looking for what doesn't exist, not what does. For example, an artisanal cheese business will have more success in lower Alabama than in the middle of Wisconsin.

Yvon Chouinard, founder of Patagonia extreme-sports clothing, feels this way, "When people copy our product and undersell us, we don't get into the fight. We either change the product or drop it. We are in the business of creating unique products, not competing." He was quoted in Robert Kriegel's *If it ain't broke...BREAK IT!*

Be weird. Business author Peter Drucker said a good judge of the validity of a niche market is to ask your neighbor what he thinks of it. If he agrees it's a good idea, it's probably not different enough to succeed as a niche market. A smaller version of a successful large business just won't cut it.

A niche fills a hole in the marketplace. It is also a hole in a person's mind. Once that hole is filled with a product's name (your company's brand) it is almost impossible for a competitor to replace it.

Authenticating your Niche. James H. Gilmore and B. Joseph Pine II writing in *Authenticity, What Consumers* Really *Want* offer these strategies for expressing your products' authenticity: Can they be more Retro or more "Un-"? Can you make them more Personal or more Premium? Will your products lead customers to focus more on Self or Other? "Develop a perspective on each...and use that to take bold new directions in bringing greater natural, original, exceptional, referential, and influential authenticity into your business."

"If you change before you have to," Sergio Zyman, author of *The End of Marketing as We Know It*, noted, "by the time the competition catches up to you, you will already be someplace else."

Change

Change. It's inevitable for growth, yet it's scary because it upsets the status quo. Change is uncomfortable because it carries with it elements of the unknown. Change very often is the root of conflict in a family business.

Once you've identified how uncomfortable change makes you, you can take the first step to overcome your squeamishness.

Benefits. Writing "Why accepting change is vital to your professional success," Connie Podista said, "Passion could be called the charge for our life's batteries. Without that charge, it's hard to get our engines revved up. That igniting charge is sparked by the challenge of change—learning new things, meeting new people, growing as professionals, and taking risks that push us to reach our potential. None of that can happen unless and until we are willing to experience the fear that inevitably arises when we move out of our comfort zones....If we want passion back in our lives, we must be willing to meet the challenge of change."

Choose to change. Acknowledge that things are going to be different. Rather than let change blindside you, become proactive. Work with the change.

Perhaps your family has decided to begin direct marketing for the first time, develop a phone call center, or even make a speech at a local business breakfast. Your old mindset says you're not good at selling or speaking in public. Why not? Who told you that? Trust yourself to meet the challenge that takes you beyond your comfort zone.

Take the test. Here's a little lesson Jim Jackson gave in *Power to Change, You Always Have a Choice.* Fold your arms over your chest. Now unfold them and do something with your hands—wave, pat your shoulder, rub your ear. Now get ready to do something differently. I'm going to ask you to fold your arms again—but wait—are your arms automatically going into the same, comfortable position as before? Reverse the position of your arms with the other on top. It feels a little awkward, doesn't it? The second time took a conscious decision on your part to overcome your natural instinct and place your arms differently. Jackson says that learning how to think differently is part of the journey to greatness.

Positive energy. Some changes happen so gradually we barely notice them, like aging or changing hair color. Others occur in crisis, which can occur so suddenly we feel helpless to react. When you make a conscious choice to change you can be the architect of the outcome. That takes positive energy, staying away from those naysayers who are always complaining. Don't participate in the coffee shop pity party. Surround yourself with positive thinkers.

The potholes in the streets of New Orleans were worse after Hurricane Katrina. One pothole was so big a refrigerator could fit in it. Complaints from people in the neighborhood didn't get anyone's attention at City Hall. One guy took positive action. He held a First Year Anniversary for the pothole and invited the media to attend the party. That got the road crew out to fix it.

Maneuver around the potholes in your life and all those skeptics who tell you something can't be done. Those are not the people to share your dream. Don't even ask them what they think. Instead practice the power of positive thinking by envisioning yourself achieving the goals you set for yourself, your family and your farm business.

Stick with those who appreciate what you're doing and can be your support group.

Be different. Products that stand out from the competition are the ones that get noticed like author Seth Godin's

Surfers rules and change

Passion rules. Successful businesses are doers. They don't just spend all their time talking about what they want to do. They demonstrate total commitment—mind, body, spirit.

No dare, no flair. They push the limits, looking for bigger waves, longer rides. They know this involves risks.

Expect to wipe out. They know that failure and mistakes are part of the learning and growth process. Trying to be too safe means never improving.

Don't turn your back on the ocean. Successful surfers have respect for the ocean knowing that if they don't pay attention they can drown or be eaten by a shark.

Keep looking at the horizon. They have to be alert and look beyond the coming big wave to see what's coming next. Change might mean the ride of a lifetime, or wipeout in the sand.

Move before it moves you. Anticipation and action is a matter of timing. Waiting too long means missing the wave and ending up in the back water. Acting too soon means being trounced by the big waves.

Never surf alone. A good network of friends and family prepares successful business people to have good backup in an emergency.

Robert J. Kriegel *If It Ain't Broke...Break It!*

purple cow in a pasture full of black and red cattle. Give yourself permission to be different too. Of all the vendors at our local farmers' market, the person we call The Goat Lady is the one who really stands out not only for her olive-and-garlic-drenched cheese, but because she's not embarrassed to promote everything related to goats.

Find something that can become your personal signature — an over-the-top hat, perky aprons, a safari theme to promote the variety of grassfed animal products in your line up. Make changes to ensure you'll be outstanding. Become a legend among other sellers as the person who always slips a poem in with every purchase or sings a line of chorus with every sale. If Allan were selling we'd blow a train whistle to create noise.

When everyone accepts you, you're no longer exciting to anyone.

Can do. There are some things no one can change—the weather, other people, and the past. Don't dwell on mistakes. Look at them as life's little learning experiences. There's a Latin saying that translated says, when the wind won't serve, take up the oars.

Everything can be improved upon. Embrace change. And don't look back.

Bibliography

Annino, Patricia, J.D., Davidow, Thomas, Ed. D. & Harrison, Cynthia Adams, Ed. D., LICSW with Davidow, Lisbeth, Ed.M., M.F.A. *Women in family business, What keeps you up at night?* Annino, Davidow & Harrison, 2009.

Attoun, Marti. "Hosting the Farmers Market." *American Profile*, July 30, 2006.

Bamburg, Jill. *Getting to Scale, Growing Your Business Without Selling Out.* Berrett-Koehler Publishers, Inc., 2006.

Barletta, Marti. *Marketing to Women, How to Increase Your Share of the orld's Largest Market.* Dearborn Trade Publishing, 2006.

Bernard, Nancy, "Sixth Annual Artists Gathering Exhbition." Ranching for Profit Newsletter, November 2005.

Blanchard, Ken & Blanz, Barbara. *The Simple Truths of* Service, *Inspired by Johnny the Bagger.* Blanchard Family Partnership, Glanz Communications, Inc., McMillan Media and Simple Truths, L.L.C., 2005

Bone, Eugenia. "Restless Palate, Dining Out," *Forbes Life,* 2009.

Brooks, Patricia. "The Country Comes Bearing Products to Go." *The New York Times*, June 29, 2008.

Burlingham, Bo. *Small Giants, Companies That Choose to Be Great Instead of Big.* Portfolio, Penguin Group, 2005.

Byczynski, Lynn. *Growing for Market* newsletter. Fairplain Publications Inc.

Cademartori, Lorraine. "Weekend in the Country." *Forbes Life.* September 5, 2008.

Cone, Steve. *Steal These Ideas! Marketing Secrets that will Make You a Star.* Bloomberg Press, 2005.

Cutler, Laura. *Inc. Magazine.*

Dreher, Rod. *Crunchy Cons, How Birkenstocked Burkeans, gun-loving organic gardeners, evangelical free-range farmers, hip homeschooling mamas, right-wing nature lovers, and their diverse tribe of countercultural conservatives plan to save America (or at least the Republican Party).* Crown Forum, 2006.

Edge, John T. "Tennessee Waltz." *Gourmet.* July 2007.

"Farm Beginnings: Heather Smith, Making ag a profitable happening." *The Land Stewardship Letter.* Winter 2010.

Ferree, Erin. "The Importance of a Logo and Marketing Materials." *allbusiness.com*

Fisher, Helen. *The First Sex, The Natural Talents of Women and How They Are Changing the World.* Ballantine Books, 1999.

Fleming, Quentin J. *Keep the Family Baggage Out of the Family Business, Avoiding the Seven Deadly Sins that Destroy Family Businesses.* Simon & Schuster, 2000.

Gilmore, James H. & Pine, B. Joseph II. *Authenticity, What Consumers Really Want.* Harvard Business School Press, 2007.

Girard, Joe. *How to Sell Yourself.* Warner Books, 1981.

Gladwell, Malcolm. *What the Dog Saw, and other adventures.* Little Brown, & Company, 2009.

Godin, Seth. *Meatball Sundae, Is Your Marketing Out of Sync?* Porfolio, The Penquin Group, 2007.

Godin, Seth, editor. *The Big Moo, Stop Trying to Be Perfect and Start Being Remarkable.* Portfolio, 2005.

Godin, Seth. *The Purple Cow.* Portfolio, 2002.

Gorman, Leon. *L.L. Bean, The Making of an American Icon.* Harvard Business School Press. 2006.

Gray, John, Ph.D. *Men are from Mars, Women are from Venus*

Groe, Isabelle. "Growers' Gamble, Are cities missing a bet by giving short shrift to farmers markets?" posted on ww.seattlefarmersmarket.org

Hamilton, Neil. *The Legal Guide for Direct Farm Marketing*

Harris, Wylie. "A Dare You Can Take...to the Bank: Alternative Farm Enterprises and Agritourism," and "Some Who Dare! To be Different." *Field Notes*, Kerr Center for Sustainable Agriculture, Winter 2005.

Hawken, Paul. *Growing a Business.* Fireside, Simon & Schuster. 1988.

Hirshberg, Meg Cadoux. "House Rules, Ten gentle suggestions for keeping the domestic seas tranquil," *Inc.* December 2009/January 2010.

Hollan, James Francis III. *The Catalog Handbook, How to produce a successful mail order catalog.* Hippocrene Books, Inc., 1991.

http://ancienthistory.about.com/od/etymology/f/Sincere.htm

http://www.maryjanesfarm.com/About/paydirt/

http://www.moneyinstructor.com/art/waysforkids.asp

http://www.moneyinstructor.com/art/childbusiness.asp

http://tyrashow.warnerbros.com/2009/03/size really_does_matter.php

http://www.utahsown.utah.gov/ConsumersloveFarmersMarkets.htm

Hyde, Julie. "Brochure Marketing: 12 Tips on How to Do it Effectively." *marketing.about.com*

Jackson, Jim. *Power to Change, You Always Have a Choice.* Action IQ Publishing. 2005.

Jaffe, Dennis T., Ph.D. *Working With The Ones You Love, Conflict Resolution & Problem Solving Strategies For A Successful Family Business.* Conari Press.1990.

James, E.W. "Dub" & Janet. *Couples at Work, How Can You Stand to Work with Your Spouse?* Boomer House Books. 1997.

Joy, Nicki & Kane-Benson, Susan. *Selling is a Woman's Game, 15 power ful reasons why women can outsell men.* Avon Books, 1994.

Keats, Nancy. "Putting with the Cows," April 25, 03. *Wall Street Journal.*

Kriegel, Robert J. *If It Ain't Broke...Break It: And other unconventional wisdom for a changing business world.* Business Plus, 1992.

Lant, Dr. Jeffrey. *Cash Copy.* JLA Publications, 1989.

Levinson, Jay Conrad. *Guerrilla Marketing Newsletter.* Guerrilla Marketing International, March-April 1994; September-October 1994; November-December 1994; January-February 1995.

Levinson, Jay Conrad. *Guerrilla Marketing Weapons: 100 Affordable Marketing Methods for Maximizing Profits from Your Small Business.* Plume, The Penguin Group,1990.

Levinson, Jay Conrad. *The Way Of The Guerrilla, Achieving success and balance as an entrepreneur in the 21st century.* Houghton Mifflin Company, 1997.

"Market Scenes: Fayetteville, Arkansas." *Growing for Market.* August, 2010.

Marshack, Kathy. *Entrepreneurial Couples, Making It Work at Work and at Home.* Davies-Black Publishing, 1998.

Michelli, Joseph A. *The Starbucks Experience, 5 Principles for Turning Ordinary into Extraordinary.* McGraw-Hill, 2007.

Monosoff, Tamara. *Secrets of Millionaire Moms.* McGraw-Hill, 2007.

Nair, Suresh. "Marketing Via Invoices and Billing Statements." *Marketing Advents*, July 2010.

Ness, Carol. "Are farmers burning out on market life?" *San Francisco Chronicle*, September 19, 2007.

Padgham, Jody. "Producer Profile: Oaks of Mamre Farm, Hempstead, Texas." *APPA GRIT!* March-April 2010.

Pease, Barbara & Allan. *Why Men Don't Listen, And Women Can't Read Maps, How We're Different & What To Do About It.* Broadway Books, Random House, 2000.

Peters, Tom. *Re-imagine! Business Excellence in a Disruptive Age.* DK, 2003.

Pine, B. Joseph II & Gilmore, James H. *The Experience Economy, Work Is Theatre & Every Business a Stage.* Harvard Business School Press, 1999.

Podesta, Connie. "Why accepting change is vital to your professional success. Http://www.agiftofinspiration.com.au/stories/success/Accepting.shtml

Popcorn, Faith. *EVEolution.* Hyperion, 2000.

Quinlan, Mary Lou. *Just Ask a Woman, Cracking the Code of What Women Want and How They Buy.* John Wiley & Sons, Inc.

Rezac, Darcy. *Work the Pond!* Prentiss Hall Press, 2005.

Royte, Elizabeth. "Weed it and reap." *Gourmet.* July 2007.

"Sales & Marketing, A virtual farmers' market, Could webcams help it grow?" *Inc.*, April 2010.

"The time team." *Farmers Journal*. July 7, 2001.

Telling, Gillian, "A Moveable Feast," *Culture, United Airlines Inflight Magazine.* November 2009.

Tingley, Judith and Robert, Lee E. *GenderSell: How to Sell to the Opposite Sex.* Second Edition. Dearborn Trade Publishing, 2006.

"'Tis the Seasaon: Selling at Farmers' Markets 101." *Animal Welfare Approved.* Summer 2009.

Ton, Vinnee. "Grocery Traps," *Times Picayune*. November 11, 2009.

Trout, Jack & Ries, Al. *Bottom Up Marketing.* Plume, 1990.

Willingham, Ron. *The Best Seller! The New Psychology of Selling and Persuading People.* Prentice-Hall, Inc., 1984.

Wright, Harold L. *Hot to Make a 1000 Mistakes in Business and Still Succeed: The Small Business Owner's Guide to Critical Deci sions.* Wright Track, May 1995.

www.AnimalWelfareApproved.org

www.brandingstrategyinsider.com

www.checkyourhealth.org

www.eatdrinkbetter.com/

www.feedstuffsfoodlink.com/

www.ksfarmersmarkets.com/blog/

www.pantone.com

www.peapod.com

www.portionplate.com

www.seattlefarmersmarkets.org

www.outstandinginthefield.com

Young, Sarah. *Jesus Calling, Enjoying Peace* in *His Presence.* Thomas Nelson, 2004.

Zaslow, Jeffrey. "Friendship for Guys (No Tears!)." *Wall Street Journal.* April 7, 2010.

Zyman, Sergio. *The End of Marketing as We Know It.* HarperBusiness, 1999.

Index

Advertising...310-318
Anger management...177-180
Artisinal markets...46
Authenticity...53-56, 353
B&Bs (see Farm stays)
Blogs...339-340
Brands...47-62
Brochures...322-334, 313
Business cards...61-62, 302-303
Buying clubs...109-110, 147-148, 245-251
Catalogs...146, 314, 325-329
Chefs...253, 259-263
Colors...200-201, 296
Community building...188-190
Database marketing...287-290
Demonstrations...88, 120-121, 262
Farm stores...235-244
Farm manual...347-350
Farm stays...274-277
Farm tours...264-273
Farmers' markets...193-212, 238, 248
Festivals...224-225, 278-284
Gift certificates & baskets...148, 230-234, 302
Giving thanks...118, 291-292
Goals...28-39, 93-94
Grassfed benefits...63-67, 101
Guarantee...67, 89, 310-311, 346
Hamburger...78
Internet marketing...333-342
Invoices...71, 343-346
Listening 115-121, 129, 131

Location...241-244, 280-281
Logos...71, 319-321
Marketing plan...93-99
Meat displays...203-206
Men & women differences...23-27, 123-139
Milestones...230-234
Mission statements...49-50
Names...68-71, 141, 305-306, 324
Newsletters...330-332
Niches...351-353
Order forms...332, 343-346
Packaging...42-43, 80, 146, 189, 224
Photographing food...204-205
Portion sizes...158
Press kits...260
Pricing...75-90, 152, 261-262
Public relations...202, 305-309, 340
Quality...42-43
Restaurants, selling to...252-253
Samples...66, 148,167-170, 197
Schools & universities, selling to...255-256
Selling tactics...115-122, 136-144, 171-176, 205-206
Shopping tips...182-183
Signage...71, 184, 197, 202
Special occasions...230-234
Story marketing...100-101, 131, 143
Surveys...340-342
Testimonials...102-103
Theatre & food...162-166, 293
Uniforms...293-297
Value added products...221-225
Working with spouse...9-22

About the author

A native Mississippian with family ties to New Orleans, I've written and photographed both states as well as much of the Southern USA. In keeping with Southern Gothic tradition, the skeletons in my closet include 14 romance novels published in 10 foreign countries, using the name, Carolyn Thornton.

Offbeat and exotic travels have sent me mule trekking in Switzerland, to a snake temple in Malaysia, following horseback fishermen in Belium, and to an Aboriginal bush camp in Australia.

I've been named Mississippi's Travel Writer of the Year. In my thirty-something years of traveling, photographing, and writing, my work has been showcased in *AAA Southern Traveler, Family Motor Coaching, Far East Traveler, Horizon, Mandarin, Mature Living, Mississippi Magazine* and *Where to Retire* among other publications.

My brother wouldn't let me play with his electric train when I was little, so I now ride trains around the world. And, thanks to Allan, we now have a working steam locomotive in our front yard.

I have worked with the *Stockman Grass Farmer* as the Green Park Press book publisher and conference manager since the early 1990s. Even earlier, when we were dating, Allan recruited me to help behind the scenes by proofreading the magazine.

Brochures For Your Customers

Are your customers confused about grassfed? Give them "Grassfed, The Informed Choice."

This 8-page digest covers the superior benefits of grassfed products. A Q&A section, plus plenty of food for thought, will guide and educate your customers and their friends in making the choice for grassfed products.

A more informed customer is a better customer.

12 copies for $15.00
24 copies for $28.00
36 copies for $40.00

Price includes shipping & handling for USA orders. All others include additional $1.50. Sorry, not sold in smaller quantities, but we can fill larger orders! Call us 1-800-748-9808 for additional pricing.

More from Green Park Press

COMEBACK FARMS, Rejuvenating soils, pastures and profits with livestock grazing management by Greg Judy. Takes up where **No Risk Ranching** left off. 280 pages. **$29.00***

FARM FRESH, Direct Marketing Meats & Milk by Allan Nation. An excellent companion to **Grassfed to Finish.** 256 pages. **$30.00***

GRASSFED TO FINISH, A production guide to Gourmet Grass-finished Beef by Allan Nation. How to create a forage chain of grasses and legumes to keep things growing year-around. 304 pages. **$33.00***

KICK THE HAY HABIT, A practical guide to year-around grazing by Jim Gerrish. How to eliminate hay - the most costly expense in your grazing operation. 224 pages. **$27.00***

More from Green Park Press

KNOWLEDGE RICH RANCHING by Allan Nation. Anyone who has profit as their goal will benefit from this book. 336 pages. **$32.00***

LAND, LIVESTOCK & LIFE, A grazier's guide to finance by Allan Nation. Shows how to separate land from a livestock business, and how to create a quality lifestyle on the farm. 224 pages. **$25.00***

MANAGEMENT-INTENSIVE GRAZING, The Grassroots of Grass Farming by Jim Gerrish. Details MiG grazing basics. 320 pages. **$31.00***

MARKETING GRASSFED PRODUCTS PROFITABLY by Carolyn Nation. From farmers' markets to farm stores and beyond, how to market grassfed meats, milk, and related products successfully. 368 pages. **$28.50***

NO RISK RANCHING, Custom Grazing on Leased Land by Greg Judy. Based on first-hand experience, Judy explains how by custom grazing on leased land he was able to pay for his entire farm and home loan within three years. 240 pages. **$28.00***

PA$TURE PROFIT$ WITH STOCKER CATTLE by Allan Nation. Profiles Gordon Hazard, who accumulated and stocked a 3000-acre grass farm solely from retained stocker profits. 192 pages **$24.95***

QUALTIY PASTURE, How to create it, manage it, and profit from it by Allan Nation. No nonsense tips to boost profits with livestock. 288 pages. **$32.50***

THE MOVING FEAST, A cultural history of the heritage foods of Southeast Mississippi by Allan Nation. How using the organic techniques from 150 years ago food crops, trees and livestock can be produced in the South today. 140 pages. **$20.00***

* All books softcover. Prices do not include shipping & handling

To order call 1-800-748-9808
or visit www.stockmangrassfarmer.com

Name ____

Address ____

City ____

State/Province ____________________Zip/Postal Code______

Phone___________________________

Quantity	Title	Price Each	Sub Total
____	**Comeback Farms** (weight 1 lb)	$29.00	________
____	**Farm Fresh** (weight 1 lb)	$30.00	________
____	**Grassfed to Finish** (weight 1 lb)	$33.00	________
____	**Kick the Hay Habit** (weight 1 lb)	$27.00	________
____	**Knowledge Rich Ranching** (weight 1 1/2 lb)	$32.00	________
____	**Land, Livestock & Life** (weight 1 lb)	$25.00	________
____	**Management-intensive Grazing** (weight 1 1/2 lb)	$31.00	________
____	**Marketing Grassfed Products Profitably** (1 1/2)	$28.50	________
____	**No Risk Ranching** (weight 1 lb)	$28.00	________
____	**Pa$ture Profit$ with Stocker Cattle** (1 lb)	$24.95	________
____	**Quality Pasture** (weight 1 1/2 lb)	$32.50	________
____	**The Moving Feast** (weight 1 lb)	$20.00	________
____	**Grassfed brochures** _____quantity_____cost @	subtotal	________
____	Free Sample copy ***Stockman Grass Farmer*** magazine		

Sub Total ________

Postage & handling ________

Mississippi Residents add 7% Sales Tax________

TOTAL ________

Shipping:	Amount	Canada
Under 2 lbs	$5.60	$8.50
2-3 lbs	$7.00	$11.95
3-4 lbs	$8.00	$15.90
4-5 lbs	$9.60	
5-6 lbs	$11.50	
6-8 lbs	$15.25	
8-10 lbs	$18.50	

Foreign Postage:
Add 40% of order.

We ship 4 lbs per package maximum outside USA.

Please make checks payable to

Stockman Grass Farmer

PO Box 2300

Ridgeland, MS 39158-2300

1-800-748-9808

or 601-853-1861

FAX 601-853-8087

www.stockmangrassfarmer.com

THE STOCKMAN GRASS FARMER

GREEN PARK PRESS